EXTENDING SCIENCE   8

LIFE WORLDWIDE
Selected Topics

Extending Science Series

Further titles are being planned and the publishers would be grateful for suggestions from teachers.

EXTENDING SCIENCE

8

LIFE WORLDWIDE

Selected Topics

Tessa Carrick MA

Stanley Thornes (Publishers) Ltd

First published in 1986 by
Stanley Thornes (Publishers) Ltd
Old Station Drive
Leckhampton
CHELTENHAM GL53 0DN
England

*An exception is made for the word puzzles on pp. 44 and 61. Teachers may photocopy a puzzle to save time for a pupil who would otherwise need to copy from his/her copy of the book. Teachers wishing to make multiple copies of a word puzzle for distribution to a class without individual copies of the book must apply to the publishers in the normal way.

British Library Cataloguing in Publication Data

Carrick, T.
    Life Worldwide: selected topics — (Extending
science; no. 8)
    1. Biology
    I. Title   II. Series
    574   QH308.7

    ISBN 0-85950-553-7

Typeset by Tech-Set, Gateshead, Tyne & Wear.
Printed and bound in Great Britain by Ebenezer Baylis & Son, Worcester.

CONTENTS

Chapter 5   **Cool Woodlands**

Chapter 6   **Polar Regions**

PREFACE

This book shows how some of the biology you learn in school applies in the world outside. You will discover how plants and animals survive in different environments. This will help you to understand more about nature conservation. I hope you will find the book both interesting and enjoyable.

The activities and questions give you a chance to study living things for yourself. Tasks include observing, experimenting and using information of different kinds. The book is suitable as a support for biology, environmental science and any general courses.

Tessa Carrick

ACKNOWLEDGEMENTS

The author and publishers wish to thank the following who provided photographs and drawings and gave permission for their reproduction:

Biofotos (cover, pp. 23, 30, 31, 39, 64 and 77)
British Museum (Natural History) (p. 72)
Bruce Coleman (cover, pp. 24, 36, 43 top and 80)
Heather Angel (p. 60)
Mrs Vivien Fifield (p. 10)
Oxford Scientific Films (pp. 27, 40, 46, 79 and 81)
Shell Photo Service (p. 52)
(The photographs on pp. 5, 6, 43 lower, 63 and 73 were taken by the author.)

ADDRESSES OF SOME WILDLIFE AND CONSERVATION GROUPS

The Nature Conservancy Council, 19/20 Belgrave Square, London SW1X 8PY.

British Trust for Conservation Volunteers, 36 St. Mary's Street, Wallingford, Oxfordshire OX10 0EU.
(Arranges work groups to look after the countryside.)

Watch, 22 The Green, Nettleham, Lincoln LN2 2NR.
(For young people who care about wildlife and the countryside.)

Wildfowl Trust, Slimbridge, Gloucestershire GL2 7BT.
(Provides grounds where wild geese and ducks feed and breed. There is a Youth Hostel at Slimbridge where you can stay to watch birds.)

World Wildlife Fund, Panda House, 11–13 Ockford Road, Godalming, Surrey GU7 1QU.
(Helps to save plants and animals all over the world.)

Young Ornithologists Club, The Lodge, Sandy, Bedfordshire SG19 2DL.
(For young birdwatchers.)

LIFE IS EVERYWHERE ON EARTH

WAYS OF SURVIVING

Almost everywhere in the world you will find something living. No matter how hard the living conditions, there will almost certainly be at least one kind of living thing (one *species*) that will survive in them. Even in hot springs, where the hot water shoots into the air as gushing geysers, bacteria and algae grow as a coloured film over the rocks. The bacteria can live at 88°C.

Some species are specialised and can cope with difficult surroundings which would kill most things. A small brine shrimp, which scientists call *Artemia*, is able to live in very salty pools, saltier than the sea, and its eggs can survive temperatures of 103°C (p. 47).

Other plants and animals are less specialised, but can survive in many places. One example is the house sparrow which you will find in towns in most parts of the world. Species which can survive all kinds of conditions may become very common.

Male (left) and female house sparrows, which are found in towns all over the world

1 cm

The water hyacinth comes from South America. Some plants have been taken to other warm countries where they have multiplied and spread. It has pretty purple flowers. As its name suggests, it always grows in wet places. In marshes it grows like the first drawing but if it is on water it will look like the second. The leaf stalks become swollen and act as floats. If the leaves are split apart, each one can grow into a new plant. It is so successful in spreading that it is now found in rivers in many warm countries. As it spreads it can block rivers and get in the way of boats.

(a)

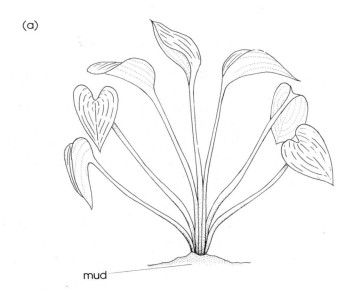

mud

Water hyacinths:
(a) growing on marsh
(b) growing on water

2 cm

(b)

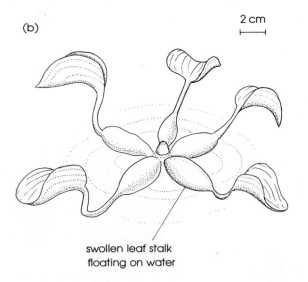

swollen leaf stalk
floating on water

In this book you will learn how plants and animals of different parts of the world survive. Each region has its own pattern of temperatures, moisture or dryness, length of day and seasons. The species which survive in each area may do so because:

- their specialised features make it possible to live there (like brine shrimps), *or*
- they can *tolerate* or put up with a wide range of conditions (like house sparrows), *or*
- they can develop in different ways in different surroundings (like water hyacinths).

Each species can live only in certain conditions. For example, a fish cannot live long on land. All plants and animals need food and space. Only species which can find enough food and space and are not eaten or killed will survive. So living things affect one another as they feed, grow and reproduce. How common each species is will depend on the conditions that the other plants and animals create.

The way in which one living thing affects another is part of the *balance of nature*. Any big change, such as cutting down trees, hunting, a volcanic eruption, or disease, will upset this balance. The numbers of each species will change and some kinds may become *extinct* (die out completely).

In each chapter of this book you can read about a region and the life in it. You will have met some of the plants and animals before. For instance, you will learn more about cacti, elephants, zebras and giraffes, camels and gerbils, polar bears and penguins. At the end of each chapter you will find some things to do and questions and puzzles to test what you have learnt. After you have read the book, visits to zoos and gardens will be more interesting. You will know more about what you see there.

CHAPTER 2

LIFE IN HOT, MOIST FORESTS

WHAT IS TROPICAL RAIN FOREST LIKE?

If you stand at the edge of a forest it will look rather like a jungle in a Tarzan film. Tarzan swings on the stems of plants clinging to the trees and pushes his way through smaller plants near the ground. The forest is like this near roads and clearings, where plenty of light can get in. In the light many plants can grow under the trees. But, deeper inside the forest it is shady and cooler. There, few plants grow close to the ground. It is easy to move about in the space between the tree trunks. The trees seem like the pillars and roof of a large cathedral or temple.

Regions of tropical rain forest

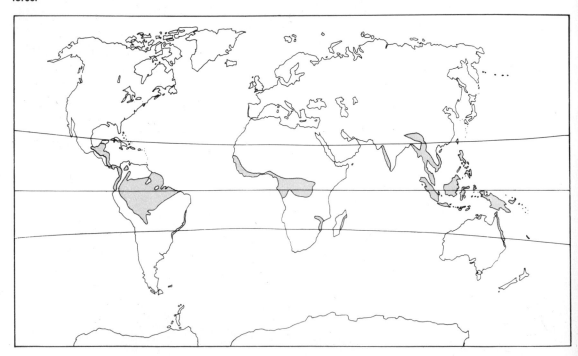

When Christopher Columbus first saw tropical rain forest after sailing across the Atlantic in 1493, he was amazed. He said the forest was 'filled with trees of a thousand kinds and tall, that they seem to touch the sky'. The average height of the tall trees is about 40 metres. A few grow to 90 metres. In Britain, a tall horse chestnut (conker) tree will grow to about 30 metres, so you can see how high the forest trees are.

Tropical rain forest from above, showing the top of the canopy

The map at the beginning of the chapter shows you where the tropical rain forest is. It is always between or very close to the tropics of Cancer and Capricorn. Days are always about the same length and there are no clear seasons. Trees lose a few leaves at a time, never all at once. They are always green or *evergreen*. Whatever time of year you visit rain forest some trees will be in flower and some will have fruit.

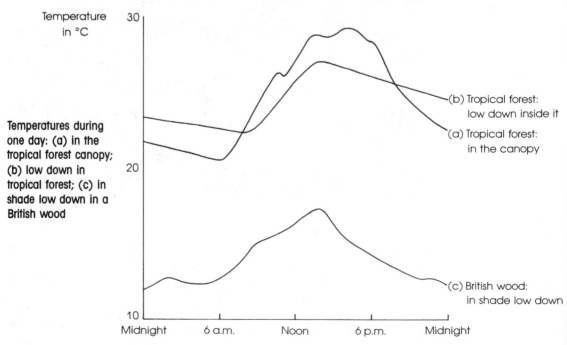

Looking into a rain forest from a path

In the forest region the rainfall is heavy, more than 160 centimetres each year and sometimes as much as several hundred centimetres. You can compare this with southern Britain where the rainfall is between 50 and 75 centimetres each year. In London there is about 60 centimetres of rain in a year, but in parts of the Lake District there is more than 250 centimetres each year. In most parts of the forest some

Temperature in °C

Temperatures during one day: (a) in the tropical forest canopy; (b) low down in tropical forest; (c) in shade low down in a British wood

30

20

10

(b) Tropical forest: low down inside it

(a) Tropical forest: in the canopy

(c) British wood: in shade low down

Midnight   6 a.m.   Noon   6 p.m.   Midnight

rain falls almost every day. There is not much change in temperature or rainfall throughout the year. The air is very humid or moist and there is very little breeze. Your skin will feel sticky and hot because sweat will not evaporate (dry up) and cool you off as it does in drier air. (See Activity 1, p. 21.)

The light is dim in the forest. If you look up into the trees you will see patches of brightness but only a very small part of the sunlight will reach where you are standing. There is a smell of earth. It is quiet except for the chirping sound of insects, some cicadas and grasshoppers, and sometimes louder cries of monkeys, apes or birds. At night it will be noisier, when the frogs and toads, more cicadas and other insects such as crickets and katydids become active and start to call.

THE LAYERS OF THE FOREST

In the forest there are enormous numbers of plants struggling to grow. Only plants which are suited to living in a particular part of the forest survive. This chapter will describe some of the ways in which plants survive in the forest.

Around you in the forest are the smooth trunks of the tall trees. The tops of the trees are arranged in layers. Layer A is the tops of the tallest trees. A few even taller ones sometimes stretch above the others. The leafy branches form the *canopy*. Next comes Layer B or the middle layer of tree tops. Below this at about 20 to 25 metres above the ground is Layer C. The diagram on p. 8 shows the layers.

Although the trees look alike from the ground there are many kinds with slight differences. One scientist trained monkeys to climb up to collect leaves, flowers and fruits for him. To find out how many different kinds of trees there are, biologists counted all large trees in 23 hectares of one forest. A hectare is 10 000 square metres. There were 2629 large trees of 381 different kinds. It is quite different in Britain where you will find only one or two kinds of large tree in a wood. In oakwoods most large trees are oaks, in beechwoods most are beeches.

One thing you will notice about some forest trees is their huge *buttresses*, up to 10 metres high, sticking out from the trunks. Also *proproots* or *stilt roots* sometimes grow down from the trunk or branches. Buttresses and proproots probably help support and anchor the tree. You can see both in the drawing of forest trees (p. 8).

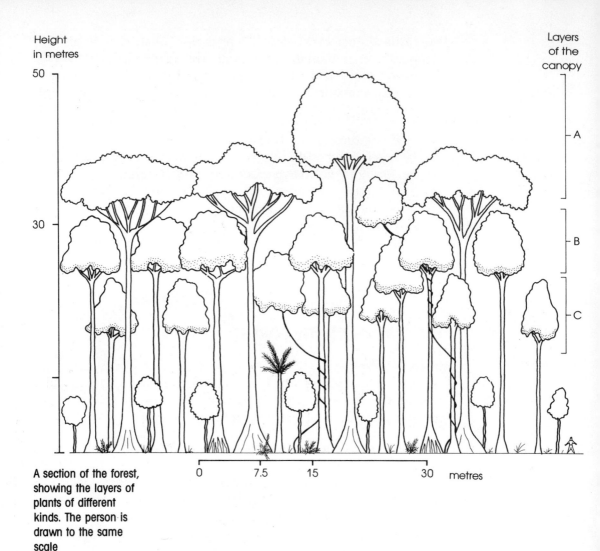

Height in metres

50

30

Layers of the canopy

A

B

C

0   7.5   15   30   metres

A section of the forest, showing the layers of plants of different kinds. The person is drawn to the same scale

Below the tall trees there are a few young trees or saplings. These saplings grow very slowly. But, if a tall tree falls the sapling puts on a sudden spurt of growth upwards to the light let into the forest.

Many people think of climbing plants when they imagine rain forest. The *lianas* that Tarzan swings on are woody, twining, climbing plants which grow up to the canopy of the forest. There are a few lianas shown in the drawing of the forest trees (above). They grow very fast, using food stored in an underground root *tuber* or swelling.

Epiphytes grow on other plants without harming them. There are many different kinds on the lower branches of the canopy. In Britain, some of these are grown indoors as pot plants.

They can survive without much water indoors where the climate is like their natural surroundings. Common South American epiphytes are urn plants or bromeliads. They grow

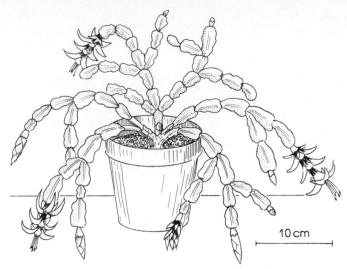

The spectacular flowers of an epiphyte, the Christmas cactus, commonly grown as a pot plant

10 cm

high up in the forest canopy. (Pineapples are bromeliad plants which grow on the ground and so are not epiphytes.) In South-East Asia strange, large stag's horn ferns grow as epiphytes. Many fantastic orchids are epiphytes, too.

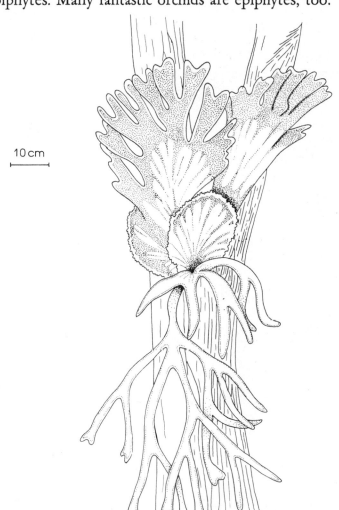

10 cm

Stag's horn fern growing on a forest tree

Orchids growing as
epiphytes on a tree
trunk (old line drawing)

The epiphytes cannot get water from the soil, but some use water caught in the cup formed by their leaves. Orchids and the climbing Swiss cheese plant have roots hanging in the air. The white tips of these roots are a mass of dead cells which act like sponges. They can soak up moisture from the air.

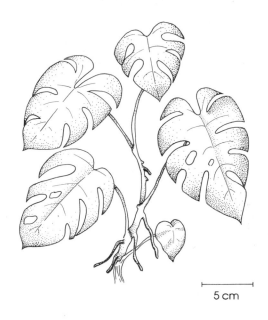

Aerial roots on a Swiss cheese plant

5 cm

SURVIVING IN THE FOREST

All forest plants compete for the things they need. Those which are able to obtain the sunlight, air, food and water for their activities are likely to survive.

HOW THE GREEN PLANTS FEED

As you know green plants do not eat but they all need food to live and grow. In fact, they make their own food from carbon dioxide and water. They need sunlight and their own green colour to help make the food. The green substance is called *chlorophyll*. The flat leaves, which are spread out in the canopy of the forest, catch the sunlight. Very little passes through to any plants growing beneath them. The carbon dioxide needed to make food passes from the air through pores or *stomata* (one is a stoma) on the leaf surface. Usually the water travels up in veins from the roots to the leaves. Making food in the sunlight is called *photosynthesis* (photo = light, synthesis = making).

11

The foods produced by photosynthesis are *carbohydrates*, chemical compounds which contain carbon, hydrogen and oxygen. Sugars and starch are examples of carbohydrates. They store the energy which came from the sunlight and so are energy-giving foods.

Photosynthesis can be shown like this:

$$\text{carbon dioxide} + \text{water} + \text{energy from sunlight} \xrightarrow{\text{green chlorophyll}} \text{carbohydrate} + \text{oxygen}$$

All the green plants in the forest make their own food by photosynthesis. Some need very bright light but plants growing below the tree canopy, including epiphytes, can use much dimmer sunlight. Plants also need small quantities of other chemicals, such as potassium nitrate. These chemicals are important for healthy growth. In most plants they come from the soil, carried with the water. In some epiphytes, chemicals are taken from the dead plants and animals trapped in the water held in the cup made by their leaves.

WATER IN PLANTS

The whole surface of each leaf is almost waterproof and the upper surface is often waxy and tough. Most of the pores (or stomata) are on the undersides of the leaves. As already mentioned, the pores allow carbon dioxide into the leaves. At the same time, water and oxygen can escape through them. The water evaporates into the air. This loss of water from the leaves is called *transpiration*. When the leaves lose water more water moves up the plant to replace it, bringing more chemicals with it.

But, if too much water is lost through the pores on the leaf the plant will suffer. The pores close when the plant begins to lose more water from its leaves than it is taking in through its roots. In the hot middle of the day the leaves of some forest trees droop or *wilt*. The pores on the lower surface are folded inside the drooping leaves away from the sunlight. This protects the plant because less water will be able to evaporate when pores are closed and the leaf is folded.

If a plant is to survive there must be a balance between four needs:

1 It needs to take in carbon dioxide for photosynthesis.

2 It needs a large surface to collect sunlight for photosynthesis.

3 It needs to lose some water because this keeps a flow of water and chemicals coming up through the plant.

4 It needs to be able to control the water in the plant when it is so hot that water evaporates very easily from the leaves.

Can you explain how the leaves help the plant to meet each of these needs?

Leaf drip-tips — what do they do?

Many tropical rain forest plants have leaves with long, tapering tips. These are called *drip-tips*.

Drip-tips on an Indian rubber plant leaf and on the leaf of a vine, *Philodendron*

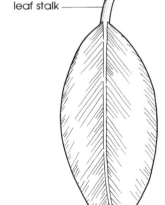

Some of the plants we keep indoors have these drip-tips. The drip-tip and the shiny leaf surface help water to fall off the leaves quickly in a rainstorm. As so many tropical plants have drip-tips it looks as if quick drainage of water off the leaf helps the plant to survive. But, no-one is sure why it is useful. Four suggestions are given below. Which suggestions do you think are most likely? Why?

1 Small plants (algae and lichens) will not grow on well-drained leaves. If they did grow they might stop light reaching the leaf and block the stomata.

2 Rain water could make the leaves very heavy. Climbing plants might fall or even pull down supporting branches.

3 If water does not fall off it will cool the plant as it evaporates. If the plant is cooled too much its inside activities will slow down and it will not grow so fast. Then, the plant will not compete very well with other plants.

4 Water on the leaves might stop transpiration (water escaping through the stomata). Then the movement of water and chemicals upwards through the plant would become too slow. As most stomata are on the underside of the leaves where rain-water could not settle, this suggestion seems rather a poor one.

Look out for the drip-tips on leaves of indoor plants. It is a clue that the plant may come from the tropical rain forest but be cautious because some other plants have them too.

ANIMALS FEEDING IN THE FOREST

All the forest animals depend on plants in one way or another for their food because only plants can trap the energy of sunlight to make their own food. Some animals feed on leaves, flowers, fruits, flower nectar or honey made from it, or on dead plant material. If they feed on plants they are called *herbivores*.

Other animals, called *carnivores*, eat animals which have fed on the plants. So, there is a *food-chain*:

plants ⟶ herbivores ⟶ carnivores

An example of a forest food-chain is:

fruit from trees ⟶ monkey ⟶ monkey-eating eagle

10 cm

Monkey-eating eagle from South-east Asia

Many kinds of animals are hidden in the forest canopy. Each has its own diet. For instance, there are eagles which eat eggs and meat, fruit-eating birds like hornbills and toucans, insect-eating flycatchers, and humming birds which feed on nectar.

Rhinoceros hornbill — a fruit-eating bird of the forest canopy in South-east Asia

10 cm

On the ground are a few *mammals* (animals which produce milk to feed their young). Most are herbivores but a few have a special animal diet of insects.

(a)

(b)

(c)

(d)

Rain forest mammals:
(a) chevrotain (Africa)
(b) tapir (Asia)
(c) capybara (South America) (d) Indian elephant (Asia)
(e) Sumatran rhinoceros (Asia)
(f) pangolin (Asia).
(a)–(e) are herbivores;
(f) eats insects.
(⊢——⊣ represents 10 centimetres.)

(e)

(f)

Besides the bigger animals on the ground there are many smaller animals, including snakes, lizards, centipedes, millipedes and scorpions. And, in one square metre of the forest floor there may be 72 700 mites, 12 000 springtails, 4000 scale insects and aphids and at least 860 ants! (Some pictures of British soil animals are on p. 66.) Nearly everything in the forest may be eaten by some animal, large or small. Imagine trying to draw all the food-chains for all the animals on one diagram. This would give a more complicated *food-web* than the one for a British wood, shown on page 65.

THE FOREST FLOOR

Most parts of forest plants can be eaten by some animal, but all the time there is a shower of leaves, fruits, branches and dead creatures dropping to the ground. Suprisingly, the ground is not covered by masses of dead material. There is not even as much dead matter as you would find on the ground in a British wood. This is because it disappears rapidly. Although there is more dead material falling it is broken down and recycled very quickly in the hot, moist conditions. Even huge tree trunks are almost gone in a year. As soon as it reaches the ground, small animals help to break up the dead material, and bacteria and fungi (moulds) begin to attack it. In the heat of the tropics these *decomposers* work very fast. All the dead material is soon used up by the bacteria and fungi. When they in turn die, useful chemicals escape back into the soil. The chemicals do not stay there long. Quickly the roots of the forest plants absorb many of them. In this way, chemicals pass from stage to stage around the cycle shown in the diagram.

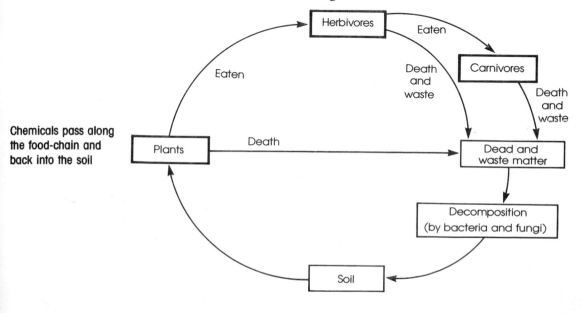

Chemicals pass along the food-chain and back into the soil

17

If they are not used most of them will be washed downwards through the soil (leached away) by the frequent heavy rain.

PARASITES

There are many parasites, — plants or animals living on and using food from other plants or animals. These could be fitted into the food-web too. One plant parasite, called by the scientific name *Rafflesia*, after its discoverer, Sir Stamford Raffles, is rare but so fantastic that it cannot be left out. In fact *Rafflesia* is one of the world's rarest plants. It grows on the forest floor of South-east Asia. It has no stem, roots, or leaves. Instead it has a few strands growing into another plant (the host) from which it absorbs food. It produces a huge, purplish, warty flower, a metre across, with five leathery petals. Its foul smell, similar to rotting meat, attracts insects to it to pollinate it. When the large flower bud opens it is supposed to hiss!

10 cm

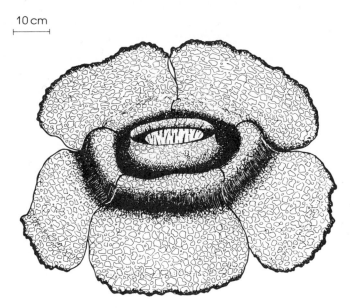

The giant flower of *Rafflesia*, a rare parasite found in South-east Asia, is sometimes called the corpse flower because of its smell

DANGER TO THE FOREST

The tropical rain forest is very complicated as well as fascinating. There is still a great deal to learn about it, quickly, before too much of it is changed, damaged or destroyed. In the forest little groups of people hunt and gather food. They clear small areas to build their homes and to plant a few crops. After a few years they move to a new site. Only a very small part of the forest is cut down and when the people move on the forest spreads back into the clearings. These forest people live in balance with the forest.

It is quite different when there is large-scale clearing of forests or *deforestation* for space for growing of crops for sale, for roads or buildings, or to sell the wood. This is a serious threat. The timber of the forest is very valuable and the money gained in selling it can be used to help tropical countries. But this leads to problems.

You can imagine how the heavy rain soon washes all goodness out of the soil after the trees are cut down. In the hot sun the bare soil often bakes hard. In other places the soil itself gets washed or blown away. The land becomes no good for farming. The forest cannot grow back once this has happened. Even if the cleared area is left the bushes and trees which will grow are not true forest. It would take hundreds, even thousands of years to regrow tropical rain forest. The complicated food-web of the forest has been upset. The delicate balance between green plants and the other living things which depend on them has been destroyed.

This does not mean that none of the wood of the forest can be used without doing permanent damage. In one scheme only a few kinds of tree are taken from the forest. Although part of the forest is damaged by this kind of *selective logging*, it is not as disastrous. The soil is harmed less and plants can grow back into the spaces. Some animals, like the gibbons, which stay in one area, may have their home areas destroyed.

The white-handed gibbon survives despite selective felling of forest trees

10 cm

Even so, some scientists have found that the numbers of gibbons stay much the same after selective logging. A few animals, for instance some kinds of frogs, may suffer if their shelter or food is destroyed.

On the other hand, the scientists found that many animals thrive in the mixture of forest and open spaces left by selective logging. Some quite rare and interesting animals do better than before the logging.

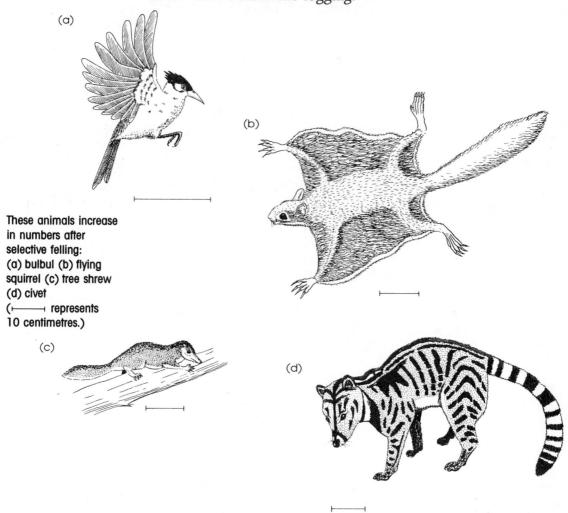

(a)

(b)

These animals increase in numbers after selective felling:
(a) bulbul (b) flying squirrel (c) tree shrew (d) civet
(⊢———⊣ represents 10 centimetres.)

(c)

(d)

In the past, some of the patches of rain forest which were left as *nature reserves* to protect wildlife were very small. They were too small to work well because they contained too few of the rarer food plants on which some animals feed. But, when selective logging is used in larger areas of forest some of the important food plants are left still growing. It looks as if selective logging provides hope for the future. Humans will be able to use some of the logs without damaging the forest soil and wildlife too badly.

SUMMARY

Tropical rain forest occurs in hot, humid conditions. It contains many kinds of large trees and other plants. These plants use the sun's energy to make food. Animals depend on some of this food. Many food-chains exist:

plant ⟶ herbivore ⟶ carnivore

Only plants and animals which are suited to their surroundings survive. Dead plant or animal material is used by bacteria and fungi which break it down, releasing chemicals back into the soil. The balance of the forest is complicated and easily upset by human activities.

ACTIVITY 1

Testing the cooling effect of evaporation on your skin

Place a drop of methanol (or methylated spirits) on the back of your hand. What do you feel? Now, place a drop on the other hand and blow it gently. Is there any difference in what you feel? Now, try to explain what happens when sweat evaporates from the skin. What happens if the air is very moist so that sweat does not evaporate?

ACTIVITY 2

Tropical plants which are kept as indoor plants

Examine some plants which come from the tropics. If you have none at home or in school try to see some in a flower shop window, a garden centre or a botanical garden. (Some plants are suggested below.)

Write notes about what you find for each plant.

Look at the leaves. Have they a leathery or waxy surface to make them more waterproof? Do they form a cup to hold water? Do they have drip-tips?

Some plants to look out for:
Bromeliads: urn plant, Amazonian zebra plant, finger-nail plant with its bright red centre.

Other epiphytes: Christmas cactus and Swiss cheese plant.

Plants with drip-tips: some *Begonia* plants, passion flower, *Philodendron*, Indian rubber plant.

21

Looking at stomata

Look for the stomata (pores) on both surfaces of a leaf of a Christmas cactus. Brush a thin layer of clear nail varnish on to a patch of the lower side of the leaf, about 3 millimetres square. When it is dry use forceps (tweezers) to peel off the layer carefully. Spread it on a clean microscope slide. Add a drop of water. Hold a coverslip at an angle to one edge of the drop.

Lower the coverslip flat, gently. Look at the nail varnish impression of the leaf surface under a microscope. You will need to use a high magnification so take care with your focusing. Look for tiny pores like the one in the drawing.

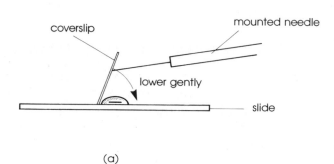

(a)

(a) Mounting a nail-varnish impression on a slide (b) The appearance of a stoma on the leaf surface of a Christmas cactus

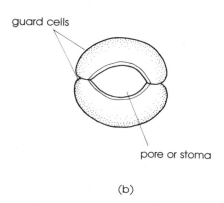

(b)

How many can you see at one time? Now do the same for the upper surface. You can try this with other leaves but sometimes it is difficult to peel off the varnish.

QUESTIONS ON CHAPTER 2

1 Look at the pictures of tropical rain forest in this chapter and make a list of the things you would not expect to see in Britain.

2 The table below shows two forest food-chains. Complete the table with the name of an animal or part of a plant in each space.

	Green plant	Herbivore	Carnivore
Food-chain 1		Monkey	
Food-chain 2	Dead leaves		Pangolin

3 In the table showing parts of a leaf, fill in what each part does.

Part of leaf	What it does
Upper surface	
Stomata	
Green colouring (chlorophyll)	
Vein	
Drip-tip	

4 Read the passages about 'Water in Plants' and 'Leaf drip-tips' (pp. 12–14) and answer the questions in the passages.

5 Make a drawing or write a poem or description to show a friend what a rain forest is like.

Flowers of an orchid from the topical rain forest

6 The photograph on p. 23 shows a beautiful forest orchid. Plant collectors search for plants like this to sell to gardeners in other parts of the world. Why do you think it would be better to collect seeds rather than whole plants?

Clearing the rain forest in Brazil, South America

7 Write down a list of suggestions to answer each of these questions:
  (a) What could the wood which is being taken from the forest be used for?
  (b) What could the cleared land be used for?
  (c) Why are some people worried about destroying forests?

HOT, DRY GRASSLANDS

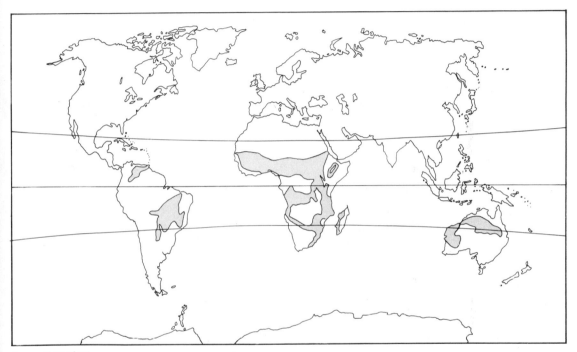

Areas of tropical grassland

A GRASSLAND NATURE RESERVE

In the drier, hot regions of the world are great belts of grassland, often called *savannas*. This chapter gives a picture of one area of savanna in East Africa, the Serengeti Plain. It is a nature reserve which is famous for its large mammals.

Tropical grassland regions have both dry and wet seasons, unlike the moist rain forests. In the Serengeti there is usually rain from about November to May but it is dry for the rest of the year. During the day it may be hotter and drier than the rain forest but at night the temperature drops quite low, down to about 15°C. In the dry season the grass is yellow but as soon as it rains new green shoots sprout. Flowers appear in the grass and birds begin to nest. The grass may grow a metre or more high.

25

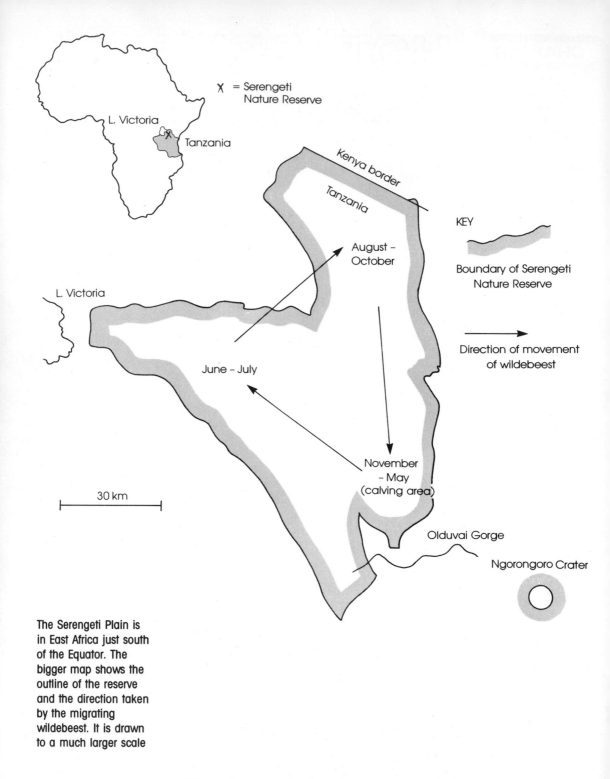

X = Serengeti
Nature Reserve

L. Victoria

Tanzania

Kenya border

Tanzania

August –
October

June – July

November
– May
(calving area)

L. Victoria

30 km

KEY

Boundary of Serengeti
Nature Reserve

Direction of movement
of wildebeest

Olduvai Gorge

Ngorongoro Crater

The Serengeti Plain is in East Africa just south of the Equator. The bigger map shows the outline of the reserve and the direction taken by the migrating wildebeest. It is drawn to a much larger scale

When the grass is dry, fires are started very easily, by lightning or people. The fires sweep across the grassland. Afterwards bright green shoots spring up making good food. Serious fires damage the soil and cause erosion. Small and young animals die but larger ones escape.

In the Serengeti there are scattered trees, up to 30 metres high. Fires and feeding animals prevent more tree growth. Flat-topped thorn trees or acacias are the most common species (see photos on pp. 30 and 31). The fire damage causes them to grow stunted and gnarled, but their thick bark protects them against the flames. Another interesting tree, the baobab, stores water in its large swollen trunk.

The Serengeti Plain has gentle hills and valleys. Here and there are small mounds of rocks where lions often rest. Bare mud hills up to 7 metres high are also found. These are built by small insects called termites which live in large colonies. Inside the hills the termites keep the air moist by smearing water from deep in the soil on to the walls of the tunnels. They feed on decomposing wood. Indigestible parts are made usable by the action of moulds growing inside the hills. The termites eat a very large amount of plant food.

Savanna grass provides food for vast numbers of herbivores – insects, birds, mice and rats, and many larger mammals. On the Serengeti Plain there are about 350 000 brindled wildebeest or gnus, more than 100 000 zebras and Thomson's gazelles, and smaller numbers of many other species. These are the largest herds of wild *grazing* (grass-eating) mammals in the world.

No wonder that the Serengeti has been made into a nature reserve where wildlife is protected. It is easy to see the animals when they visit water-holes to drink.

Zebras and wildebeest at a water-hole. The zebras' stripes make it difficult to pick out single animals

The herbivores eat grass and leaves. In their turn they provide food for carnivores, including large mammals like lions, and birds such as vultures. Essential chemicals pass along the food-chain and back into the soil in an endless cycle. (See the diagram on p. 17.)

The large mammals are bothered by flies and parasites. Sometimes you will see small oxpecker birds sitting on an animal, feeding on the parasites. The biting tsetse fly causes particular distress. It sucks the blood of many mammals, including humans, and spreads a microscopic parasite called a trypanosome from animal to animal. The trypanosome gives people sleeping sickness. When another tsetse fly bites the sick person it sucks up more parasites to pass on to its next victim. Tsetse flies and sleeping sickness have prevented people from settling in the Serengeti Plain.

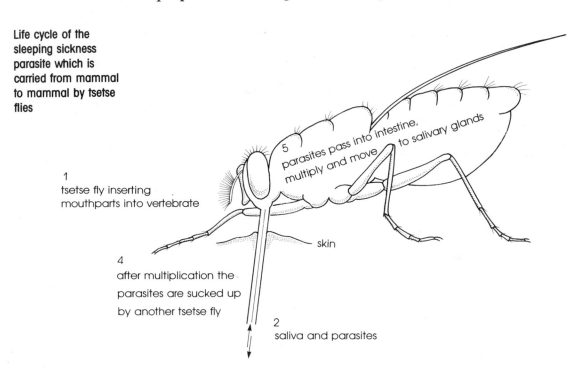

Life cycle of the sleeping sickness parasite which is carried from mammal to mammal by tsetse flies

5 parasites pass into intestine, multiply and move to salivary glands

1 tsetse fly inserting mouthparts into vertebrate

skin

4 after multiplication the parasites are sucked up by another tsetse fly

2 saliva and parasites

3 parasites pass into vertebrate blood, multiply, and may cause sleeping sickness

GRASS — THE FOOD OF THE HERBIVORES

On the plains, where it is much drier than in the rain forest, trees do not survive well. Grass grows instead. It makes its food by photosynthesis (see p. 11). When the grass plants are trampled or eaten or burnt, new leaves grow up again from buds at the soil surface. This is like the new growth after mowing a lawn. Only drought stops the grass from growing. The wildebeest eat nearly all of the grass leaves and move on, but a month later the plants are back to normal. The new leaves contain plenty of useful protein and sugars and so make good food for grazing herbivores. Even though the grass makes good food, herbivores have to feed nearly all the time to get enough to keep them fit.

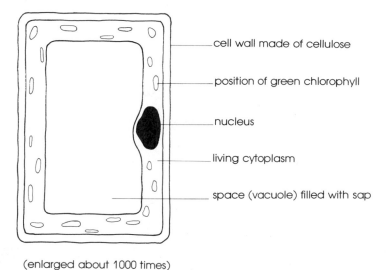

cell wall made of cellulose

position of green chlorophyll

nucleus

living cytoplasm

space (vacuole) filled with sap

One cell from a grass plant showing its tough cell wall

(enlarged about 1000 times)

Grass cell walls are made of tough, indigestible cellulose. Herbivores' teeth are suited to this tough plant food. The front *incisor* teeth are sharp and can bite off food. Behind these is a gap or *diastema* in which the rough tongue can move the food about. The back teeth have sharp ridges which grind up the cell walls. These ridges wear away as the animal gets older and can be used to tell how old a herbivore is, for example, you can tell a horse's age from the wear on its teeth.

In a herbivore's *alimentary canal* (digestive tube) there is often a part where bacteria and one-celled animals (or protozoa) live. These bacteria and protozoa can digest the cellulose walls of the plant cells. Wildebeest and cows belong to a group of mammals called *ruminants*. They have a special part of the stomach (the *rumen*) which holds these bacteria and protozoa. After the food has been mixed up with the bacteria and protozoa and partly digested it is brought back into the mouth for the animal to chew it again. This is called 'chewing the cud'. The food is then swallowed once more and the useful part is absorbed.

THE MIGRATING HERBIVORES

If you go to the Serengeti at the right time of year you will see very large groups of zebras, wildebeest or Thomson's gazelles, all moving together. They *migrate* along much the same path every year, travelling about 300 kilometres (see map on p. 26).

The zebras come first, in herds made up of families. Each family has one male, several females and their young. They eat the tougher, longer grass. As they have no rumens for bacteria to digest the food, they have to eat even more grass to get enough goodness.

Migrating wildebeest streaming across the savanna. A few acacia trees can be seen

Next come the wildebeest. They often walk in single file, so animals in front warn the others of any lurking lion, waiting in ambush to attack. The wildebeest eat the long leaves left by the zebras and crop the grass nearly to the ground. As they feed they look about, watching for danger. Their eyes, set high on each side of the head, allow them to see all round. When one wildebeest senses danger the others quickly respond and the herd moves off.

Wildebeest mate while the herd is on the move. A male ready to mate or *in rut* collects about ten females about him. He drives away other males. He prances around the females, grunting and bellowing, and mates with any females ready for him. Males are in rut for about two weeks and during that time feed very little. Later, the females have their calves in the grassy area marked on the map (p. 26). The young are ready to run with the herd very quickly. When very small there is a risk they will be eaten or have an accident. But there are so many young born at once that some survive.

Finally, the daintier Thomson's gazelles move along the migration route, feeding on the short new grass which grows up after the wildebeest have finished. As they feed the gazelles watch for danger. When they are anxious they do a peculiar jump which startles enemies. In this *stotting* movement they leap in the air with all four legs held out stiffly and their tails upright. If a carnivore comes too close the gazelles dart away, distracting enemies by taking a zigzag path.

At the driest season the migrating animals are in the moister western or northern parts of the Serengeti. As soon as rain comes again they set off back to the east, to feed on the newly growing grass. There would not be enough food for them if they were to stay in one place. The long lines of animals moving together is one of the most impressive sights in the world. The three kinds of migrating herbivores live well together. Each uses those parts of the grass left by the others.

OTHER HERBIVORES

On the plains there are many other plant-feeders, such as buffaloes, large elands, and pig-like warthogs. Some of these feed in groups where they warn each other of danger. They move around but do not migrate along the same routes each year.

Giraffes feeding on an
acacia tree

Giraffes wander over a small region, feeding or browsing on leaves of trees. A giraffe can stretch up to 6 metres. If seeds of the acacia trees are eaten and pass through a giraffe's alimentary canal they will develop or *germinate* more easily into new plants. So, giraffes prevent trees from growing very tall by their feeding but help more small trees to grow. In this way they provide more food for themselves in the future.

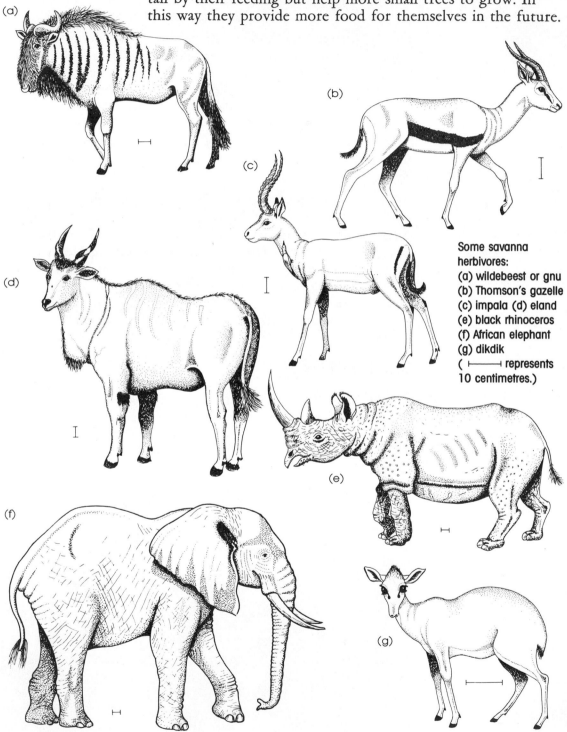

Some savanna herbivores:
(a) wildebeest or gnu
(b) Thomson's gazelle
(c) impala (d) eland
(e) black rhinoceros
(f) African elephant
(g) dikdik
(├────┤ represents 10 centimetres.)

Impala feed on grass and also browse on bushes. They too help acacia seeds to germinate. They live in herds of 10 to 100 animals and stay in a fairly small area. Each breeding male has a *territory* or patch of land which it marks with a smelly substance coming from its forehead. They roar and fight with challenging males as they protect their territories. Females are attracted into these territories to mate. When they are ready to give birth, females leave the herd. They return when the young can keep up with the moving adults.

In the wooded areas you will find tiny dikdik, weighing only 4 kilograms. Like the impala, dikdik have territories. They move around as families of one pair and the young. The family territory is marked, this time with a smelly substance from glands near their eyes and with piles of dung. Other dikdik families are warned off and when the young are old enough they are driven away to find their own territories. The territory contains a mixture of kinds of food plant.

Each kind of herbivore has a different way of getting enough food. Some migrate regularly to new grass, some wander about to find their food, and some defend territories where there is enough growing to meet their needs. Usually, they all find enough food to survive and their different habits mean they do not compete too much. But, in the dry season some die. About one out of every ten wildebeest dies of starvation in a very dry year. Fortunately, there are plenty more wildebeest to keep the species going. Usually, it is the healthiest animals which survive and produce the next generation.

The African elephant — a question of size

African elephants are bigger and have much larger ears than Indian elephants. An elephant weighs about 1000 times more than a tiny dikdik. Each day it needs about 160 kilograms of grass, leaves and bark, so it has to feed most of the time. It must have a large area to explore for food, water and shade.

Very large mammals have problems in keeping cool. Their big bodies produce a lot of heat. But, they have less skin surface for each cubic centimetre of body volume than a tiny animal like the dikdik has. You can understand this if you look at two Plasticine blocks. The small one has sides 1 cm long and a volume of 1 cm^3.

The large one has sides 10 cm long and a volume of 1000 cm³.

So the length of the large cube is ten times the length of the small one.

Cubes of 1 cm side and 10 cm side

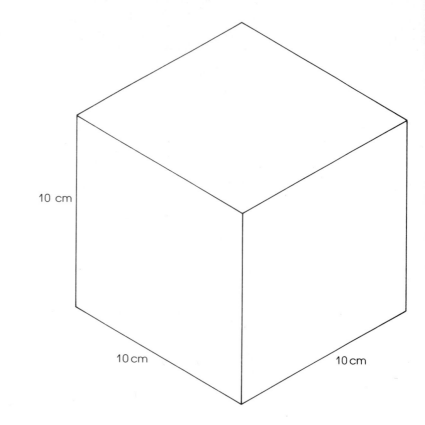

Its volume is 1000 times the volume of the small one. Its mass will also be 1000 times that of the small cube, just as the elephant is 1000 times heavier than the dikdik.

But what about the area of the surface?
For the small cube, the area of each side is

$$1 \text{ cm} \times 1 \text{ cm} = 1 \text{ cm}^2$$

There are 6 sides so the whole surface area will be

$$6 \times 1 \text{ cm}^2 = 6 \text{ cm}^2$$

For the large cube, the area of each side is

$$10 \text{ cm} \times 10 \text{ cm} = 100 \text{ cm}^2$$

For the 6 sides of the large cube, the total surface area will be

$$6 \times 100 \text{ cm}^2 = 600 \text{ cm}^2$$

So, the small cube has 6 cm² of surface for 1 cm³ of volume and the large cube has 600 cm² of surface for 1000 cm³ of volume.

The large cube has only one tenth as much surface for each cubic centimetre as the small cube.

So, in a huge animal like an elephant there is not much surface from which it can lose all the heat its body makes. The African elephant which has survived in the grasslands can cope with this problem. It has very few hairs on its skin. Hairs would keep the heat in. It has very large ears which increase its surface and can be flapped to help lose heat to the air. And, whenever it can, it bathes in water or mud. When the moisture dries up by evaporation it cools the body (see Activity 1, p. 21). This is one reason why the elephants need a good water supply.

The black rhinoceros — will it survive?

These rhinos are fairly shy, browsing mammals found in wooded parts of the plain. They use old paths over and over again. Once there were a great many black rhinos but so many have been shot that they have become scarce.

Local farmers sometimes shoot rhinos to stop them walking across their land. But, most of the killing is done as a sport or to get rhino horn. The horn has more value than gold in parts of Asia where it is sold as a special medicine. Even though shooting rhinos is against the law, it is still difficult to stop it all. If the number of rhinos becomes too low, very few young will be produced to carry on the species.

In the Serengeti, shooting has killed most black rhinoceroses. The *Red Data Book*, which lists all rare animals, says the black rhino is *vulnerable*. If nothing is done to protect it, the rhino could become even more rare or *endangered*. Endangered species easily die out or become *extinct*. The black rhino is not the only vulnerable species in the Serengeti. The table on p. 38 lists some rare carnivores.

Most people believe that rare animals and plants should be saved for the future. In order to do this, as much as possible must be found out about the animal or plant and its *habitat*.

The habitat is the place where it lives. Scientists working in nature reserves and in zoos are learning more about how each animal lives and breeds. Sometimes, it is even possible to send some of the animals bred in zoos back into the wild. *Conservation* is the protection of wildlife and wildlife habitats.

SERENGETI CARNIVORES: PREDATORS AND SCAVENGERS

Most of the carnivores of the Serengeti are well known.

African hunting dogs eating their prey. Although their spots make them difficult to see, their tails can act as signals

There are fewer of them than there are herbivores. Some large carnivores only eat meat which they have caught themselves. They are *predators*. Others, *scavengers*, feed on animals which die or which have been killed by other carnivores. The table on p. 38 shows you which carnivores are predators, scavengers or both. Each kind of carnivore hunts at different times and feeds on different prey. In this way they avoid too much competition for food.

15cm

Some savanna
carnivores: (a) cheetah
(b) leopard (c) hyaena

	Lion	Leopard	Cheetah	Hunting dog	Spotted hyaena
Mass	100–200 kg	35—60 kg	40–60 kg	17–20 kg	45–60 kg
Numbers in Serengeti	Reasonable (about 2000)	Reasonable (about 1000) Vulnerable species	Rare (about 300) Vulnerable species	Rare (about 200) Vulnerable species	Reasonable (about 4000)
Habitat	Grass and trees	Trees	Grass and trees	Grass and trees	Grass
Hunting time	Mainly at night	Night	Day	Day	Night and dawn
Hunting group size	1–5	Alone	Alone	1–3 or 4–20	Pack (2–19)
Food (prey)	Zebra, wildebeest, buffalo, Thomson's gazelles, warthog	Impala, Thomson's gazelles, dikdik, reedbuck	Thomson's gazelles, Grant's gazelles, impala	Thomson's gazelles, wildebeest, zebra	Wildebeest calves, Thomson's gazelles, zebra
Hunting style	Stalks to 100 m; then chases; strangles prey	Stalks prey	Chases prey, up to 110 km/h (70 mph); strangles prey	Chases for up to 3 km	Chase herd in group or scavenge
Predator	Yes	Yes	Yes	Yes	Yes
Scavenger	Yes	Occcasionally	No	No	Yes

In the Serengeti there are also predatory birds, such as eagles, and many bird scavengers, including vultures.

Griffon Vulture

These glide overhead until they see a dead animal below. Then they drop down to the food to tear it apart with their strong hooked beaks and feet. Often when lions and hyaenas see vultures fly down they set out to share the food. Hyaenas also follow the migrating wildebeest, zebras and Thomson's gazelles and feed on the weakest animals.

Most people find lions exciting. In the Serengeti, you can often see a group or *pride* of them resting together in the open. There will be several females and their young, with perhaps one or two males. Females (lionesses) stay together, in the same pride, all their lives but the males may leave. The pride keeps its territory for very many years.

A lion pride at rest.

For part of the night the lions hunt. Meat provides more food than grass and so carnivores do not need to eat all the time to get enough. Although they do not always hunt together food is often shared between them. When they hunt, lions stalk up to their prey, ambush it and catch it in a short distance. Their powerful jaw muscles and sharp teeth, particularly the long *canines*, strangle and tear the prey apart. Hyaenas have even more powerful jaws and can crush bones.

The male lion's open mouth shows its pointed, meat-eating teeth

CAMOUFLAGE

Camouflage makes an animal look like something else, so that it is hidden in its surroundings. Many Serengeti animals show some kind of camouflage. Lions and many young herbivores are a golden colour which matches the dry, yellow grass when they lie still. Usually, an animal which matches its background has a pale underside which overcomes the way in which shade makes it look darker. Other animals have patterns, like the giraffe's markings and the spots of the leopard, cheetah and hunting dogs. These help them look invisible in patches of light and shade. Sometimes, an animal which blends well into its background has one part of its body which stands out so that it can signal to other members of the group. The tails of Thomson's gazelles and of hunting dogs and the black backs of lions' ears show up in this way.

The camouflage of some animals confuses predators. For example, when you are near a zebra its stripes make it stand out clearly. But, if you see a lot of zebras together in the hazy distance, as a hunting lion does, it is difficult to tell where one ends and the next begins. This makes it hard to pick out one animal to stalk.

SUMMARY

In tropical grasslands, grass and tree leaves are important plant foods. The herbivores have different ways of obtaining enough food. They may migrate along a regular route, wander over wide areas, or hold a feeding territory. Territory may be defended for breeding as well as for food. There are fewer carnivores than herbivores. The carnivores include predators and scavengers. The large mammals of the Serengeti Plain are suited in many different ways to survive there. Examples are the way the elephant keeps cool, the lion's meat-eating teeth, and the use of camouflage.

The Serengeti nature reserve helps to protect wildlife and allows scientists to study it. A few of the animals are rare and if not protected could become extinct.

ACTIVITY 4

Looking at teeth

Look at the teeth, skull and lower jaws of a herbivore such as a rabbit and a carnivore such as a dog. The numbers and arrangement of the teeth will be different but the ways in which the teeth are shaped will be like those of African mammals which eat similar food. Examine the shape and cutting edges of the teeth. Draw or write notes to show how the teeth are suited to a plant or a meat diet.

ACTIVITY 5

Camouflage

1) Make a copy of the imaginary animal and cut it out. Colour or shade it so that it is camouflaged against the grass pattern on the right.

2) Look for animals or pictures of animals which have a darker back than underside. Make your own coloured drawings. How does each animal's colouring help to camouflage it?

QUESTIONS ON CHAPTER 3

1 Make five columns with the headings shown below. List the mammals named in the chapter under the correct headings. Some will belong in more than one column, some will not fit in any of them.

Migrating herbivores	Tree and shrub feeders	Carnivores which hunt by day	Carnivores which hunt by night	Vulnerable mammals

2 Describe in your own words what the migration of wildebeest across the Serengeti looks like. Use the picture on p. 30 to help you. How does this migration help the animals to survive?

3 Fill in the spaces in the following sentences. Do not write on this page.
Some mammals, like the tiny ___ live in small family groups. Each ___ has a patch of land or ___ which it guards and where it can find all the ___ it needs. It marks its ___ with a ___ substance and drives away other families.

4 Choose one of the carnivores from the table on page 38 and describe in your own words how it feeds.

African elephant bathing

5 Examine the photographs showing an African elephant, tropical oxen, and the hippotamuses on the cover. Make a list for each of them of all the ways you can suggest in which they are able to keep cool in very hot surroundings.

Hippopotamus	African elephant	Oxen

Oxen from the tropics

CROSSWORD ON HOT, DRY GRASSLANDS

First trace this grid on to a piece of paper (or photocopy this page — teachers, please see note at the front of the book). Then fill in the answers. Do not write on this page.

Across

1 They catch their own prey (9)
4 There are few ___ on the Serengeti Plain because of sleeping sickness (3)
6 A scavenging bird (7)
9 A wildebeest (3)
10 Browsers feed on these (one only) (4)
11 She lives in a pride (7)
12 Very fast carnivore (7)
15 Large herbivore (5)
16 These could become extinct (6)
18 Indigestible material of plant cell walls (9)

Down

2 All animals need to do this (3)
3 Fly which carries dreaded disease (6)
5 Birds produce them after rains begin (4)
6 Such an animal could become extinct says the *Red Data Book* (10)
7 Lone spotted hunter among the trees at night (7)
8 Acacia ___ germinate after the giraffe has eaten them (5)
13 Tearing tooth of carnivore (6)
14 Main plant food in the Serengeti (5)
15 High on the wildebeest's head so they can see all around (4)
17 All the energy in the plant's food comes from this (3)

SURVIVAL IN THE DESERT

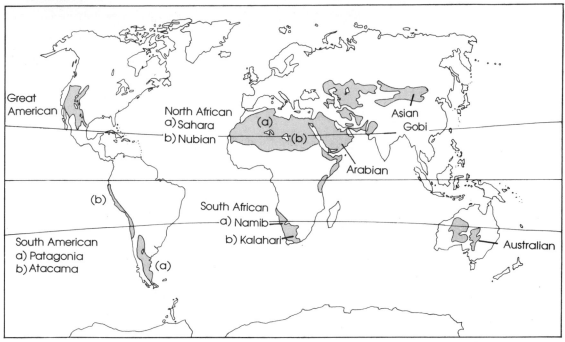

Desert regions

WHAT IS A DESERT?

The map shows you some of the very dry or *arid* regions of the world. All these places have high average temperatures and little rainfall. No-one can tell when rain will come and some years there may be almost none. The rain which does fall runs over the surface and evaporates away immediately. The soil is very dry. Any moisture there is dries up leaving behind dissolved chemicals, so some desert soils are quite salty. True deserts have an average annual temperature of more than 18°C. During the day the temperature sometimes rises to above 40°C. At night it will be much cooler.

Desert soil is usually shallow, with very little dead plant material or *humus*. In some places the sand blows up into dunes. In the shelter of dunes, dead plant and animal matter collects and provides food for animals. Hard rock which has not worn away stands up above the sandy or stony earth, leaving lumps which are so impressive that one place in the Great American Desert is called Monument Valley.

45

Desert is hot, sunny and dry with very few plants

If you stand in the desert in the middle of the day it is hot, dry, perhaps windy, with no shelter except some bare rocks. All around you it is golden brown. If the wind is strong the sand will fly into your eyes. You will feel hot, sweaty and thirsty. At night the temperature will drop and then you will feel cold.

It is difficult to find enough to eat and drink in a desert. Humans survive as nomads moving from place to place, keeping a few animals to graze on any plants there are. In some places there are *oases* where rocks containing water come to the surface. Around the water, plants grow but if the area is used too much these plants die and sand may blow over the oasis and destroy it. Some rivers, like the Nile, flow from wetter regions through deserts. Plants which grow on their banks make a green strip across the golden desert.

Any plant or animal which survives in the desert has to be able to tolerate the harsh conditions or to escape from them. This chapter shows some of the ways in which living things exist despite the challenging surroundings.

After rain

For a short time after heavy rain the dry land changes. Pools of water collect. In them all kinds of plants and creatures appear, as if from nowhere. In fact, they have grown from seeds, spores or eggs which have been *dormant* (inactive) since the last rain. As soon as rain comes the eggs hatch. The young develop rapidly, reproduce and form new dormant stages before the pool dries up again.

One kind of animal which appears in temporary, salty ponds is the brine shrimp. You can watch the life history of the brine shrimp if you follow the instructions in Activity 6 (page 58).

(a)

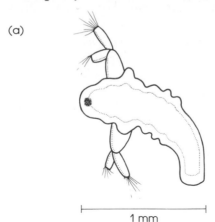

1 mm

Brine shrimps:
(a) young (b) adult

(b)

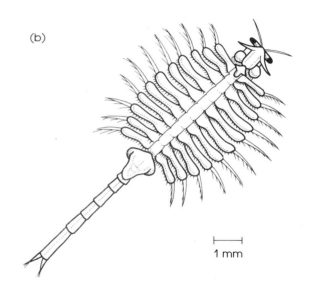

1 mm

The flowering of the desert

In the same way, after rain some plants grow up quickly. Most of the time these plants are inactive seeds, able to stand high temperatures up to 100°C and very dry conditions. Some of these seeds contain chemicals in their outer coats which prevent germination. The chemicals are washed away if there is enough rain. Then, the seeds germinate and grow rapidly, soon producing bright flowers. Suddenly the arid land becomes full of colour. The flowers produce fruits which contain new seeds. Hot, dry conditions are ideal for ripening and drying the seeds so they are quickly ready to survive until the next rain.

The Californian poppy, from dry regions of North America, grows well in dry, sunny parts of our gardens (Activity 7). The stem and leaves are covered by a bluish wax which prevents the plant losing too much water. Bright orange flowers are produced after about two months and these attract insects. The insects spread pollen from flower to flower (*pollination*). After pollination, each female cell is fertilised and develops into a seed. Each fruit holds many small seeds which are soon ready to be scattered.

Plants which pass the driest periods as resting or dormant seeds escape some of the problems of desert life. They are called *drought evaders*.

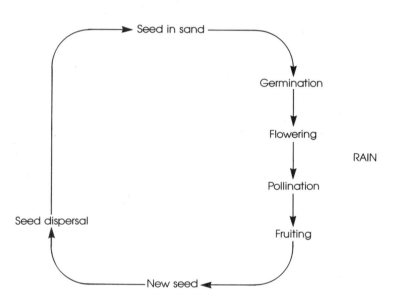

Life cycle of a drought-evader plant

TOLERANT PLANTS

Other plants that grow in deserts *tolerate* (put up with) difficult surroundings. Plants able to do this are called *xerophytes* (meaning dry plants). They may be able to recover after drying up or they may have special features which enable them to cope with the conditions.

Succulent plants

Succulents are fleshy plants like the cacti which everybody thinks of growing in deserts. In fact some of the best known deserts like the Sahara in North Africa have very few succulent plants. But in the Great American Desert and the South African Desert, plants swollen with stored water are common. One famous cactus is the giant American saguaro which grows slowly to a height of several metres. Cacti may contain more than 95% water (Activity 8) so you might think they would make a good water supply for humans and other animals in the desert. However, their sap is usually poisonous or tastes unpleasant. A few, like the prickly pear, have parts which can be eaten.

Cacti have special features which enable them to survive by protecting them from losing water. If you could measure the area of the surface of a Saguaro cactus and compare it with the surface of a leafy British tree of the same height, you would find that the surface of the cactus is much smaller. It has a very regular shape and no flat leaves. The surface is waxy, making it waterproof, and its shininess reflects some sunlight away. The spines are all a cactus has in the way of leaves. (These also protect the plant from feeding animals.) The number of stomata in the *cuticle* (tough surface layer) is less than in many other plants. Finally, the surface of the swollen stem is pleated or grooved so that one part shades another and the airflow over the surface of the plant is low. All these specialisations reduce the amount of water lost in the hot, dry climate. The roots, spreading over a fairly large area, collect what moisture there is.

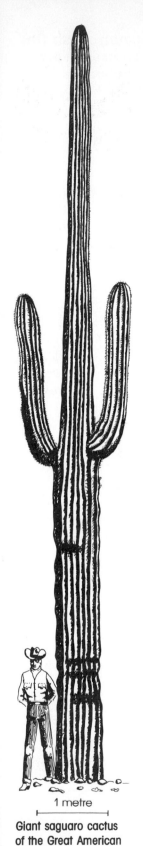

1 metre

Giant saguaro cactus of the Great American Desert

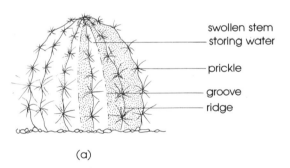

swollen stem storing water

prickle

groove
ridge

(a)

Approximately natural size

(b)

Cacti: (a) a spherical cactus with a small surface compared with its volume (b) a cactus with flattened stem held parallel to the sun's rays

49

In cacti and some other succulent plants, photosynthesis (the method of food production) is different from most other plants. In Chapter 2 (p. 11) you learnt about photosynthesis. The process was summarised as:

$$\text{carbon dioxide} + \text{water} + \text{energy from sunlight} \xrightarrow{\text{green chlorophyll}} \text{carbohydrate} + \text{oxygen}$$

The carbon dioxide needed for photosynthesis enters the leaves of most plants during the day, through the open stomata. Water can escape through the same stomata.

In a cactus the stomata behave differently. They are closed during the day and open at night, when it is cooler and less water will be lost. Once night arrives carbon dioxide diffuses (or moves) into the leaf through the open stomata and combines temporarily into compounds. When it becomes light these compounds release the carbon dioxide inside the leaf and it then takes part in producing food by photosynthesis in the sunlight. The oxygen leaves the leaf at night. The result in the end is the same but the timing of the opening of the stomata keeps down the loss of water by transpiration.

Another interesting succulent is the South African candle plant. The plant's surface is covered by waterproof, bluish wax. It has fleshy stems which store water and carry out photosynthesis. When there is plenty of water the plant also has leaves. As soon as water becomes short these leaves wither and die, cutting down the area which loses water. When water returns, new leaves grow on the younger stems (Activity 9).

(a) Candle plant with fleshy leaves after frequent watering (b) candle plant without water

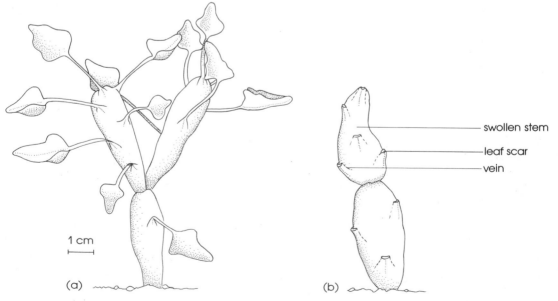

1 cm

(a)

swollen stem

leaf scar

vein

(b)

Like many other plants the candle plant reproduces in two ways. If a piece of stem is broken off it can produce roots and become a new plant. The plant also produces yellow dandelion-like flower heads, which, after pollination, form feathery fruits like the 'parachutes' of dandelion clocks. The fruits blow away so that new plants can grow away from the parent.

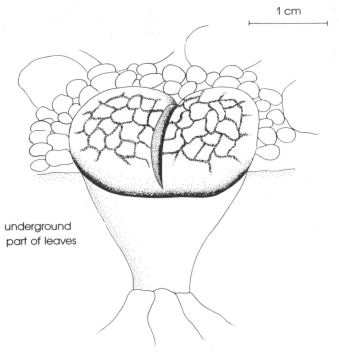

1 cm

Stone plant, showing surface of leaves among the pebbles and the underground swollen base of the leaves

underground part of leaves

Stone plants also grow in the South African Desert (Activity 7). They are buried in the stony soil. Only the tops of their leaves can be seen, matching the pebbles around them. Although they are so well camouflaged, baboons and bustards (a large bird) eat them.

Desert plants are unusual in many ways and they may turn out to be of use to humans. It is important that they should be protected. For this reason the World Wildlife Fund is keeping a careful watch on rare plants. There are strict laws controlling export of rare plants and animals from country to country.

DESERT LOCUSTS AND OTHER SMALL CREATURES

In arid regions few bacteria exist. Instead, scavenging (p. 36) invertebrates (animals without backbones) break up and decompose animal corpses and plant remains, releasing materials back into the environment for re-use. One-celled protozoa, snails, woodlice, beetles and other insects all help.

Part of a locust swarm

The destructive locusts eat growing plants. A quotation from the Old Testament of the Bible reminds us of how long these pests have been a nuisance.

> 'At his command came locusts, hoppers past all number, they consumed every green thing in the land, consumed all the produce of the soil.'

(Psalm 105: 34–35.)

Locusts look like grasshoppers, but behave differently. When they become abundant they change to what is called the *gregarious phase*. Their behaviour alters and they collect together in swarms of millions of insects. The swarms migrate and can spread over thousands of square kilometres. Helped by the wind, they travel 50 to 100 kilometres in a day. Wherever they descend they eat the plants. Their powerful external jaws bite off pieces and their forelegs guide these into their mouths. Crop losses can be enormous.

Although the desert locust can exist in dry regions, the females lay their eggs in moist sand. In the gregarious phase, the locusts place their egg pods close together. When the eggs hatch the young stages or *hoppers* feed, moult and grow. As soon as the new generation become winged adults, the migration continues.

Records of locust movements have been kept for more than one hundred years. Scientists at the locust section of the United Nations Food and Agricultural Organisation (FAO) in Rome study locusts carefully. Satellite pictures, radar and local records are all used to forecast swarm movements so that crops can be protected by chemical insecticides. Fortunately, in very dry periods even the locusts are affected by the drought and their numbers drop. Those that remain belong to the *solitary phase* and no longer swarm or migrate.

MAMMALS IN THE DESERT

A desert animal can only survive if it can prevent itself from drying up. In hot dry climates there is a great risk of losing too much water. The table below lists several ways in which mammals can gain or lose water:

Water gains	*Water losses*
Drinking From water in food Chemical breakdown of food to give water	Through surface In faeces (unused food) In urine From lungs

Heat is another problem. Although it is usually very hot, desert mammals do not have enough water to be able to use it as sweat for cooling. They survive because they have other ways of coping.

There are three common methods making it possible for mammals to survive in deserts. Some, like the small Dorcas gazelles, move about to find the places where there is some water. A second group shelter during the day. The third way in which animals survive is to be able to tolerate heat and dryness. Camels fit into this group.

Burrowing jerboas, kangaroo rats and gerbils

All these mammals are *nocturnal* or active at night when it is cooler and the air is moister. They escape into burrows during the day. Typical conditions above ground and in the burrow are shown in the drawing of North African jerboas.

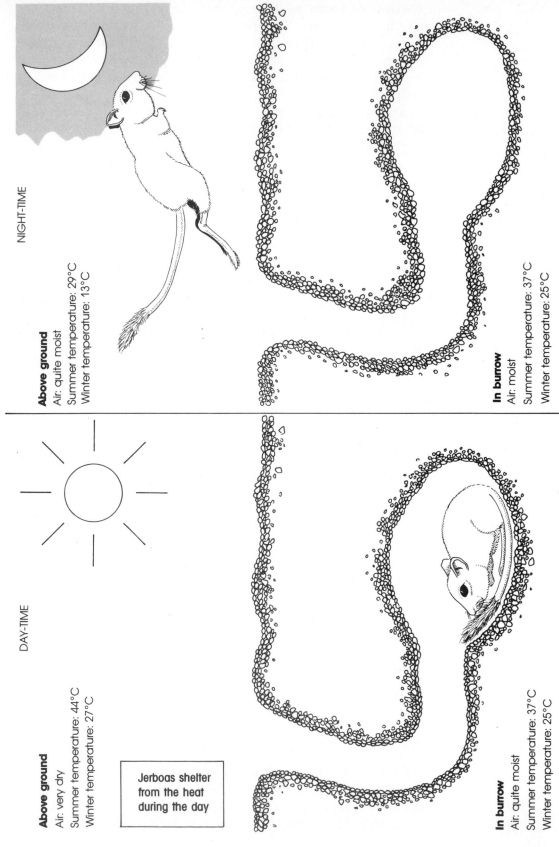

NIGHT-TIME

Above ground
Air: quite moist
Summer temperature: 29°C
Winter temperature: 13°C

In burrow
Air: moist
Summer temperature: 37°C
Winter temperature: 25°C

DAY-TIME

Above ground
Air: very dry
Summer temperature: 44°C
Winter temperature: 27°C

Jerboas shelter
from the heat
during the day

In burrow
Air: quite moist
Summer temperature: 37°C
Winter temperature: 25°C

(a)

⊢ 1 cm ⊣

(b)

⊢ 1 cm ⊣

Large ears are common in desert mammals
(a) Fennec fox from North Africa (b) Kit fox from North America

Although the jerboas of North Africa and the kangaroo rats of North America are not related they are very alike in looks and behaviour. Over millions of years the animals which have survived in the two desert areas have done so because of similar features. Here are some of the ways in which they are similar:

- Their main food is dry seeds.
- Water is gained by breaking down of food chemicals inside the body.
- The faeces are dry.
- They produce very little urine.
- They do not sweat.
- They are active at night.
- They burrow during the day.
- Their strong hind legs and front legs help in burrowing.
- Their large ears, feet and tail give a large surface for losing heat.
- Their large eyes and ears detect danger in the open desert.
- Their sandy colour helps to camouflage them.

Gerbils (Activity 10) also live in arid lands where they shelter in burrows during the day. Some kinds of gerbil make good pets, which are easy to keep clean because they produce dry faeces and very little urine. Like jerboas, they feed mainly on seeds.

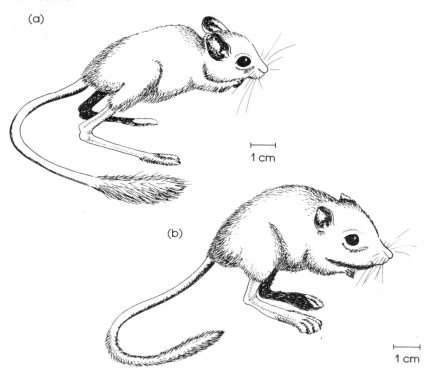

(a)

⊢ 1 cm ⊣

(b)

⊢ 1 cm ⊣

Similar mammals from the North African and Great American Deserts.
(a) jerboa from North Africa
(b) kangaroo rat from North America

Camels

An old joke says, 'The camel is an animal designed by a committee'. In order to fit in all their ideas the committee members came up with a very odd-looking animal. But camels can survive where there are few other living things and they make it possible for humans to live, travel and work in desert regions. The one-humped dromedaries from Africa and Arabia are domesticated or kept for humans to use them. A few of the two-humped Bactrian camels in Asia are wild, but most of them are kept by humans too. The most noticeable features of camels are their humps which are stores of fat, a reserve of food. Because the fat is all in one place, rather than spread out under the whole skin, it allows heat to escape from the rest of the body's surface.

Dromedary, with detail of foot

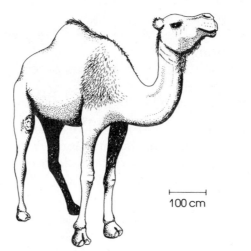

100 cm

Toes spread out to help movement over loose sand

Camels are suited to desert life in many ways. Some of the most important are:

- They can go without water for long periods.
- They can drink water very fast when they have a chance to do so.
- They can tolerate their body temperature rising during the day as long as they can cool off at night.

THE SPREADING DESERT

True desert is interesting and beautiful. But, when desert spreads to land which was once fertile it is worrying. In 1880 less than one-tenth of the land on Earth was arid. By 1950 about a quarter was desert-like. Travellers in the nineteenth century describe woodland where now there is desert. An ugly word, *desertification*, is sometimes used for the ugly spread of dry, barren land.

The causes of desertification are complicated. On the border between desert and savanna the balance is very easily upset. If the number of people increases they need more food and more firewood to cook it. Cutting down trees, damage by fire, too much farming, grazing and trampling, all lead to poorer soils. Once soil is no longer protected by grass and trees it can be washed away by sudden rain or be blown away by wind. If there are several dry years the damage done by overuse will be more severe. The desert will spread, wildlife will die and farming will produce less. One of the results can be terrible famine when hundreds die each day.

It is very difficult to stop desertification. Channelling water (*irrigation*) to crops has to be done very carefully to avoid more damage to the soil. It can also take water from other places where it is needed. Also, the water can become a breeding place for water snails which spread a disease called bilharzia, blackflies carrying river blindness, and mosquitoes spreading malaria.

The story below is a hopeful one, taken from the magazine *New Scientist* in 1984. Hasaniya district is near the Nile in the Sudan. There has been no rain for five years. The wandering nomads, with their goats and sheep, are in difficulties. During drought the few desert plants have been eaten by the animals, leaving bare soil. This soil has dried and blown away.

A forester, working for an organisation called 'Green Deserts', is trying to prevent the soil erosion. He is planting a special kind of tree called the mesquite. The trees are difficult to start growing, but when they are a few years old they begin to stop wind erosion. They have bacteria in their roots which help to trap or fix nitrogen from the air in the soil and so make the trees grow better. The trees do not need watering, and they produce nourishing seed pods, shade and firewood. Now that local families have seen the usefulness of the trees, they are asking for more to grow themselves. In the long run the 'Green Deserts' project may be more useful than food in helping to overcome the problems caused by the spread of arid lands in Africa.

SUMMARY

Desert plants and animals have many special features which enable them to survive very hot, dry conditions. Some avoid the worst periods. Others are able to cope with water shortage and great heat.

Desertification, the spread of arid regions, is a serious problem which is not easily solved.

Brine shrimp (*Artemia*)

Add a pinch of dry brine shrimp eggs to a 10–15 per cent salt solution (about 1 dessertspoonful of salt in one cup of water). The eggs can be bought from a tropical fish shop. Keep in a warm place and top up the water level with distilled water when necessary.

Examine some dry eggs under a microscope. Next day, examine some soaked eggs. After 2 days there should be young stages and these can be fed by adding a tiny amount of yeast in a few drops of water. You can watch the brine shrimps develop into transparent adults over a month or more (see illustration on p. 47).

If your eggs do not hatch try again with new eggs or alter the conditions slightly.

Growing seeds

Plant seeds of Californian poppy (*Eschscholtzia*), in a sunny garden or in a pot on a sunny window-sill.

Also, plant some cactus and stone plant (*Lithops*) seeds in trays of sandy, gravelly soil.

Watch the plants grow and develop.

How much water is there in a cactus?

Use a balance to find the mass of an evaporating dish. Cut open a small cactus, place it on the dish and find the new mass. Keep overnight in an oven at 90–100°C. Cool and find the mass again. Repeat the heating until no more mass is lost. Work out the percentage of water in the cactus.

Mass of dish (A) $\qquad = \qquad$ g

Mass of dish + cactus (B) $\qquad = \qquad$ g

Mass of dish + cactus after final
   heating (C) $\qquad = \qquad$ g

Original mass of cactus $\qquad = (B - A)$ g

Mass of water in cactus $\qquad = (B - C)$ g

Percentage water in cactus $\qquad = \dfrac{(B - C) \times 100}{(B - A)}$ %

Candle plant and water

Examine a healthy candle plant (*Senecio articulatus* or *Kleinia articulata*) and record all the ways in which it is suited to desert life.

Keep two healthy plants in identical conditions, with good light. Water one plant twice a week. Water the other plant after every two weeks. Record any differences in the appearance of the two plants after a month. Then, find out what happens if both plants are watered twice a week.

ACTIVITY 10

Mongolian gerbil

Mongolian gerbil

2 cm

Examine a gerbil and list the ways in which it is suited to life in arid conditions.

QUESTIONS ON CHAPTER 4

1 Imagine you have been dropped from a plane into a desert. You are rescued after 24 hours. Write down what you think you would see and feel during your time alone in the desert.

2 Make a table using the headings given below. Fill as many spaces as possible with names of plants and animals which survive the difficulties of desert life in each way.

	Plants	*Animals*
Resting stages or dormancy		
Cutting down on water loss		
Storing water		
Taking in water quickly		
Special behaviour		

3 Brine shrimps (*Artemia*) used to live in brine pools at salt-works in Britain. Explain how they were suited to life in these unusual habitats.

4 Explain how large ears may help North African fennec foxes and North American kit foxes to survive in the desert.

Mesquite tree in
natural habitat in USA

Leaves and flowers of
the mesquite tree

5 Read p. 57 about the usefulness of mesquite trees in deserts. Pretend you are the leader of a group of desert people. Write a speech which you will use to try to persuade the rest of your group that it will be useful to grow mesquite trees on your land.

WORDFINDER ON SURVIVING IN THE DESERT

Copy the grid below and then find the answers to the clues. One has been ringed. There are 25 more words to find. The words read from right to left or left to right horizontally, from top to bottom or bottom to top vertically, or diagonally. You can use the same letter more than once. Do not write on this page.

1 Resting stage (7)
2 Large desert mammal (5)
3 Feed (3)
4 Dry (4)
5 Wet area in a desert (5)
6 African burrower (6)
7 On a camel's back (4)
8 Salty shrimp (5)
9 Swarming locusts are ____ (10)
10 A wandering person (5)
11 ____ helps to make things hard to see (10)
12 A succulent plant (6)
13 Desert mammals often have large ____s (3)
14 A drought ____ avoids the heat and dryness (6)
15 A desert tree (6)
16 Without water (3)
17 Useful for looking around (4)
18 Has only one of number 7 (9)
19 A pore on the leaf surface (5)
20 A hot, dry region (6)
21 Shortage of food (6)
22 Dead body (6)
23 Blows up into dunes (4)
24 A plant living in dry places (9)
25 A kind of fox (6)
26 It germinates (4)

G	U	A	N	D	O	R	M	A	N	T
R	A	C	O	R	P	S	E	M	U	R
E	C	A	M	O	U	F	L	A	G	E
G	S	C	A	M	E	L	D	E	E	S
A	R	I	D	E	R	E	D	A	V	E
R	B	A	S	D	A	V	O	G	E	D
I	R	L	F	A	M	I	N	E	L	J
O	I	X	E	R	O	P	H	Y	T	E
U	N	A	N	Y	T	H	U	E	U	R
S	E	D	N	A	S	R	M	S	T	B
E	A	R	E	T	A	E	P	S	M	O
I	R	Y	C	A	C	T	U	S	P	A

COOL WOODLANDS

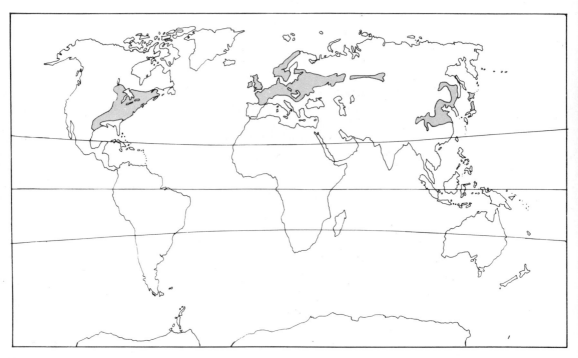

Temperate
woodlands

WOODS IN BRITAIN

Britain has a *temperate* climate, with sufficient rainfall for plants to grow well, temperatures which are usually neither too hot nor too cold, and seasons. Most of Britain would be covered by woods if left alone (see photograph, p. 63), but farming and buildings have replaced the woodland. In England and Wales the main trees are *deciduous*, with leaves which fall in the autumn. In Scotland, natural woods contain evergreen pines. Only along coasts, on high mountains, and in some rocky or very damp places do trees find it hard to grow.

Young trees
growing on a
disused railway
track

WOODLAND PLANTS

The kinds of trees growing in a wood depend on the soil and
other conditions. Where the main trees are oaks, they reach
about 10 metres high. Gaps between the leaves and spaces
where old trees have died let light through, allowing seedlings

Tree seedlings:
(a) oak growing
from an acorn
(b) beech

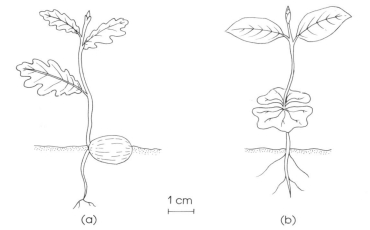

1 cm

(a)   (b)

and smaller trees and shrubs, like elder and brambles, to grow
(see diagram, p. 64). Grass and flowering plants form a field
layer and on the floor of the wood are mosses and fungi.

A beechwood in spring as the leaves open. The leaves give such deep shade that the ground below the trees is bare

In a beech wood, the leaves in the canopy cut out so much light that few things grow beneath them. The floor of the wood is almost empty once the leaves have appeared but some plants do grow and flower in the spring. Light is very important for all green plants and affects where and how they grow (Activities 11 and 12).

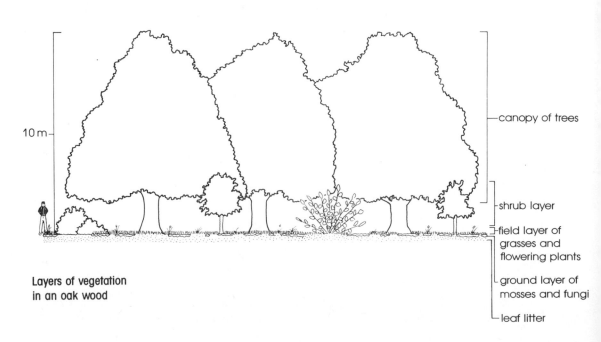

10 m

canopy of trees

shrub layer

field layer of grasses and flowering plants

ground layer of mosses and fungi

leaf litter

Layers of vegetation in an oak wood

As in other parts of the world, green plants produce food by photosynthesis, using the sunlight which falls on them (p. 11). Plants are at the beginning of every food-chain as food for herbivores (plant-eaters), such as aphids (greenfly). Herbivores are fed on by carnivores (animal-eaters), perhaps great tits, which may be eaten by a large carnivore like a sparrow hawk. Many woodland food-chains are put together in the food-web below:

Part of a woodland food-web

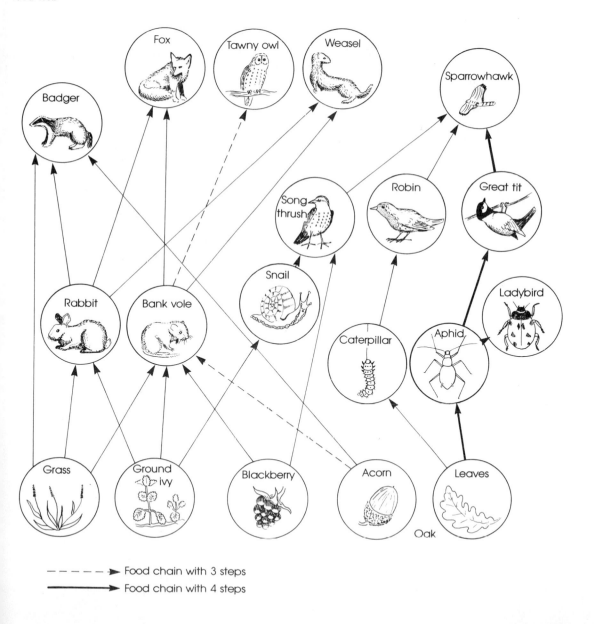

- - - - → Food chain with 3 steps
——→ Food chain with 4 steps

DECOMPOSITION

Falling leaves settle on the ground. Gradually, they are broken up by invertebrates and decomposed (rotted) by fungi. In one square metre of soil there are hundreds of different kinds of living things. There may be several million bacteria and more than a kilometre of threads of fungi, millions of nematodes (tiny threadworms) and protozoa (one-celled animals), and thousands of mites and springtails.

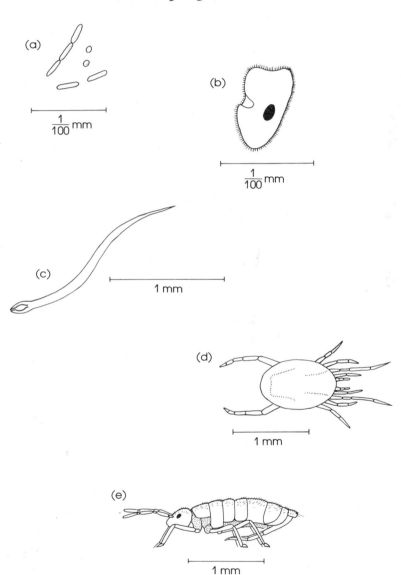

(a)

$\frac{1}{100}$ mm

(b)

$\frac{1}{100}$ mm

(c)

1 mm

(d)

1 mm

(e)

1 mm

Some soil organisms:
(a) bacteria
(b) a protozoan
(c) a nematode or threadworm from the water-film in the soil
(d) a mite
(e) a springtail

Compare this with the rain forest soil animals (p. 17). All these living things help to recycle chemicals in the dead materials on the woodland floor. The diagram on the next page shows the recycling of carbon.

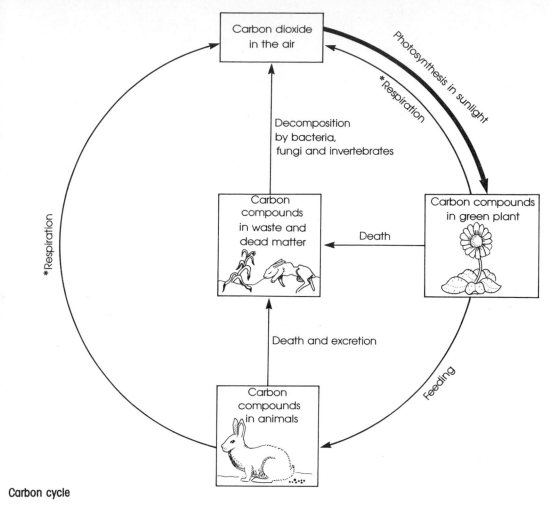

Carbon cycle

*Respiration takes place all the time in all living cells. Respiration uses carbon compounds to produce energy for all activities. Carbon dioxide gas is produced at the same time.

SEASONS IN THE WOOD

In spring many woodland plants flower before the trees' leaves appear. Later, in the leaves' shade, small plants grow slowly. Some, like bluebells, die down when they have scattered their seeds, remaining only as underground bulbs. During the summer, the trees' twigs grow in length and become thicker, using food produced by photosynthesis. Young leaves are eaten by insects, but older leaves contain unpleasant chemicals which protect them from many herbivores. The trees flower and produce fruits.

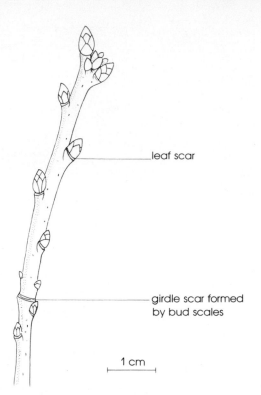

Scars on an oak twig

leaf scar

girdle scar formed by bud scales

1 cm

Autumn brings shorter days and colder weather. Deciduous trees lose their leaves before frosts and winds can damage them. Before leaves fall, useful materials move from them into the twigs. Then, a waterproof layer develops across the base of each leaf. Eventually, the veins to the leaf are blocked and it snaps off, leaving a waterproof scar on the twig. Above each leaf scar is a tiny, dormant bud. The tree stays inactive over the winter. Next spring the dormant buds produce new, leafy shoots.

WOODLAND ANIMALS

Just as the plants change with the seasons, so animals' activities alter. The hordes of insects living on trees grow rapidly in early summer and reproduce while there is food for them. When conditions worsen most insects become inactive, often spending the winter in a resting state called *diapause*, under tree bark or as eggs or pupae. Many animals find winter as difficult to survive as a dry period is in a desert.

In spring there is an invasion of small birds which have migrated to Britain from warmer wintering regions. While the weather is mild, they nest and find plenty to eat. Once breeding is over, in early autumn, these birds return south. Other kinds of birds stay in Britain all year. These are mostly seed or fruit eaters or are carnivores, feeding on other birds and small mammals, and they find enough food for the species to survive.

In the woods shrews, mice and voles survive the winter underground or in holes. Squirrels build up stores of nuts and shelter much of the winter. Bats and hedgehogs *hibernate*. The food hibernators eat in late summer is stored as a special brown fat in their bodies. At the beginning of autumn, they hide and become sluggish. They become *cold-blooded*, like invertebrates and frogs. This means that their temperatures go up as their surroundings get warmer, or down when the conditions get cooler. Their heart beat and breathing rate become much lower. Gradually, their brown fat is used up. Only when days get longer and warmer in spring do bats and hedgehogs become active again and their body temperatures return to normal.

SUMMARY

Woodland plants and animals are able to survive together because their feeding needs are different. Seasons are very important to them. Most growth and reproduction happens during spring and summer. In autumn and winter they are less active. They may become dormant, shelter or hibernate, or avoid cold and lack of food by migrating to other countries.

ACTIVITY 11

A woodland clearing

Find a gap between the trees in a wood. Write a list of the plants in the lighter clearing which are not growing under the shade of the trees.

ACTIVITY 12

Bluebell study

1) When bluebells are in flower measure the height of ten plants growing in a shaded place and count the number of flowers on each plant. Record your results in a table. Then do the same for ten plants growing in a light place. Find the average height and number of flowers for plants in shade and in light. (Find the average by adding up the results for your ten plants and dividing by ten.)

	SHADE		LIGHT	
Plant	*Height/cm*	*Number of flowers*	*Height/cm*	*Number of flowers*
1				
2				
etc.				
Average				

2) Count the number of seeds in one ripe fruit from each of ten bluebell plants growing in shade and ten plants in light. Record your results in a table and find the averages for shade and for light. After counting the seeds scatter them in the region where you collected them. Do plants growing in shade or in light have the highest number of seeds?

ACTIVITY 13 — Animals in a wood

Collecting apparatus:
(a) using a pooter to collect insects
(b) detailed view of pooter (c) pitfall trap (d) funnel for collecting animals from soil air
(e) funnel for collecting animals from water-film in soil

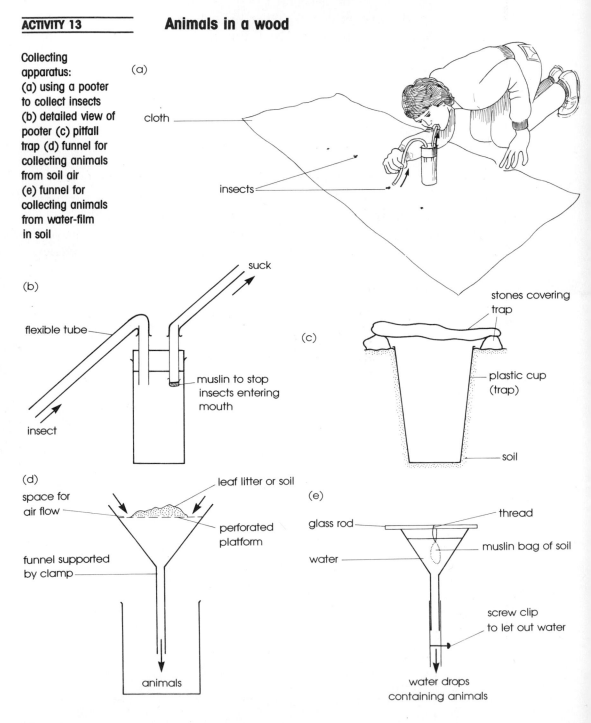

(a)

cloth

insects

(b)

suck

flexible tube

muslin to stop insects entering mouth

insect

(c)

stones covering trap

plastic cup (trap)

soil

(d)

space for air flow

leaf litter or soil

perforated platform

funnel supported by clamp

animals

(e)

glass rod

water

thread

muslin bag of soil

screw clip to let out water

water drops containing animals

1) Hold a cloth under a branch of a tree. Shake the branch and catch what falls off on the cloth. Fold the cloth over to trap the animals. Use a pooter to suck up each animal so you can examine it. Try to count how many kinds of animals you find.

2) Place pitfall traps in the ground (see drawing c). Visit them regularly to examine what has fallen into them.

3) Set up funnels like those in drawings d and e. Place a light 20 cm above them. Use a microscope to look at the animals falling into the container of funnel d and in a sample of water from funnel e.

QUESTIONS ON CHAPTER 5

1 Copy the table below into your notebook and use the food-web on page 65 to help you complete it. One example has been done for you. From the food-web you can see that the robin feeds on caterpillars so it is a carnivore. It is food for a sparrow hawk. (Omnivores feed on both plants and animals.)

Plant or animal	Position in food-web	Feeds on	Is food for
Robin	Carnivore	Caterpillars	Sparrow hawk
Grass	Producer	————	
Rabbit			
Hedgehog	Omnivore		
Songthrush	Omnivore		
Great tit			
Sparrow hawk			————
Fox			————

2 Suggest how you would try to find out the positions of plants and animals in the food web in a wood, hedgerow or garden.

3 Discuss the following:
The local paper says that there are plans to build a main road through Great Bluebell Wood. Do you think this matters? Why? What can you do about it?

4 The drawing shows four steps in a food-chain. Fill in the table to show where each living things gets energy from? What happens to the energy in each case?

	Where energy comes from	*What happens to energy*
Oak Leaf *Caterpillar* *Shrew* *Badger*		

A wood in
the spring

5 Examine the photograph of a wood in the spring. What
do the plants on the woodland floor need to enable them
to grow? What differences in the woodland will there be
in the summer?

CHAPTER 6

POLAR REGIONS

Arctic

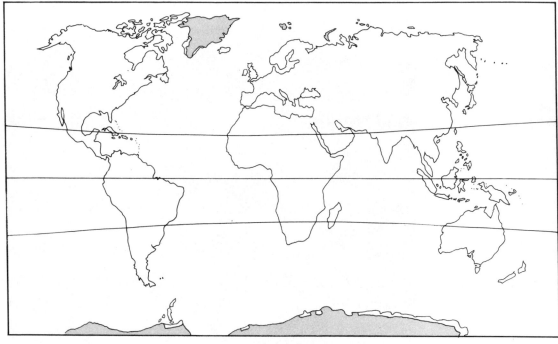

Polar regions

Antarctic

THE ICY POLES

On the map you can see the icy polar regions. There is no land at the north pole, only frozen sea. At the south pole, the hilly land is covered by ice. Along its edges the ice breaks up, forming pack ice. Ice floes and icebergs float away from the main mass. As the ice breaks it creaks and crashes noisily.

During winter months there is almost continuous darkness, but in summer there are long light days. The skies are grey but the ice appears a wonderful blue. Each year only a few centimetres of snow fall. Because of the bitter cold this does not thaw much, although sometimes water collects on the surface of the ice. Fierce, howling winds often rage over the ice.

74

In the sea, during the long summer days, microscopic floating plants, *phytoplankton*, multiply rapidly, using the light for photosynthesis. At the edge of the southern Antarctic region, called the Atlantic Convergence, currents bring useful chemicals or minerals up from the seabed and this makes the phytoplankton grow particularly well.

THE ANTARCTIC FOOD-WEB

Phytoplankton is the plant food on which all the animals depend. In the Antarctic, phytoplankton is eaten by *krill* rather like shrimps. (Scientists call krill *Euphausia superba*!) Krill is unevenly spread through the sea. It forms great *superswarms*, several kilometres across and up to 10 metres deep. Krill can be eaten by humans, but if much were caught the larger animals of the Antarctic would lose their main food supply. Vast numbers of carnivorous animals depend on krill for food. In turn these carnivores provide food for bigger carnivores, which are called *secondary carnivores* because they are a second level of carnivores in the food-web.

Part of the Antarctic food-web showing the importance of krill

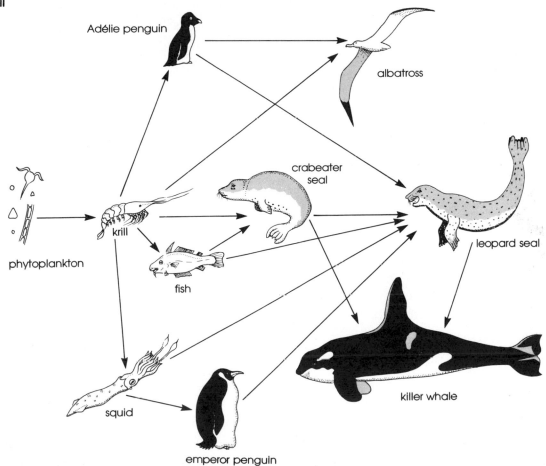

The drawing of part of a food-web and the table of seals below show you some of the animals that could be harmed if people began to use krill for human food. In the past whales and fur seals have been hunted for oil, fur and other useful materials. This, too, affected the food-web. At last, some control on whaling has begun, and, in 1982, the Convention for the Conservation of Antarctic Marine Living Resources (CCAMLR) began to protect all life of the Antarctic seas.

Antarctic seals

There are six kinds of seals in the Antarctic. Their breeding sites, behaviour and food differ so they do not compete greatly. They fit into the wildlife of the Antarctic in different ways — they occupy different *niches*. From the table you can work out several ways in which the six kinds of seals differ.

Seal	Breeding site	Approx. numbers (1982)	Size	Special features	Food
Crabeater	Edge of ice	30 million	2–2$\frac{1}{2}$ m	Sparse hair, blubber	Krill, some fish and squid
Elephant	Land	700 000	Male 6 m Female 3 m	Sparse hair, blubber	Fish, squid
Fur	Land	900 000	1$\frac{1}{2}$–2 m	Dense fur	Krill, fish, squid
Leopard	Edge of ice	250 000–500 000	3–4 m	Sparse hair, blubber	Penguins, seals, krill, fish, squid
Ross	Pack ice	250 000–500 000	3$\frac{1}{2}$ m	Sparse hair, blubber	Mainly squid, some krill and fish
Weddell	Pack ice. Winters under ice. Dives deeply.	750 000	3$\frac{1}{2}$ m	Sparse hair, blubber	Fish, squid, bottom invertebrates

Seals are mammals which are well suited to sea life. The fur seal has a dense insulating coat, which used to attract hunters. The others have hair which gives a smooth surface to their streamlined bodies. All the seals except the fur seal are insulated by a thick layer of *blubber* or fat under their skins. Their internal temperature is kept at about 37°C despite the icy outside conditions.

Weddell seals can survive under the ice, breathing at air holes and diving to 600 metres. Under water, they probably use sounds to find their way, by using echoes of the sounds they make. This is called *echo-location*. They mate in water, but young Weddell seals are born on the ice. Crabeater seals (see food-web drawing, p. 75) dive more shallowly, feeding on krill at night when it is near the surface. They can live as long as forty years. Fur seals also feed at night, travelling over an area of 250 kilometres.

Penguins

Seventeen different kinds of penguins live in the southern hemisphere, but there are none living wild north of the equator. Usually, the further north they breed, the smaller the penguins are (see diagram).

Some penguins of
the southern
hemisphere

Name of penguin	Emperor	King	Adélie	Gentoo	Chin strap	Jackass	Galapagos
Latitude South at which it breeds	70°	45–70°	70°	45–70°	53–70°	34°	0°

South Pole ◄————————————————————► Equator

As you learnt on page 34, large animals have less surface for each cubic centimetre of their body. This means that large warm-blooded birds and mammals have relatively less surface through which they lose heat and so are well suited to cold conditions. There is not an exact fit between size and distance from the equator. Other factors, such as variations in climate, behaviour, and special adaptations are also important. But, on the whole, penguins show a trend towards larger size in colder climates.

The largest penguin, the emperor, lives in the Antarctic and lays its eggs in the bitterly cold southern winter. Adélie penguins also breed very far south, but not in winter. The emperor penguin's year is described below.

March	Penguins return to breeding site (rookery)
April	16 hours darkness; temperatures down to −35°C. Penguins 'sing' and court partners.
Late May	One 12 cm egg laid by each female. Male places egg under fold of skin, balanced on top of his feet. Female returns to the sea to feed on squid.
June-July	The male protects the egg through winter. All the males huddle together. They do not feed but use up some of their blubber. They are so well insulated by blubber, feathers and other penguins, that they sometimes eat snow to keep themselves cool.
End of July	Egg hatches. The chick is nearly naked at first. Female returns to feed young on partly digested fish. Male returns to the sea to feed.
August to November	Parents go off to feed in turn. By September both may be away together, fetching more food for the young bird.
December	Young birds are carried away on the breaking ice. They moult before they enter the sea to feed. All the birds are now at sea.

The emperor penguins' rigid pattern of behaviour allows the species to reproduce in the most difficult conditions.

Polar bears

A mother polar bear and her cub wandering across the Arctic ice

To the north, in the Arctic, polar bears roam over kilometres of ice, hunting for food. There are several different kinds of seals living in the Arctic, all different from those of the Antarctic. Ringed seals are very common and are an important food of the polar bear. Bears often wait beside a breathing hole in the ice, until a ringed seal comes up to breathe. Then they kill it quickly. Arctic foxes may stand around nearby to pick up the remains.

Polar bears are well suited to their cold habitat. They are some of the largest of the world's bears, although Canadian big brown bears are larger. Compared with the small tropical Malayan sun bear, both these bears of cold regions are, like the big penguins, less likely to lose heat. Polar bears have small ears and hairy feet. Under their thick, oily fur and their skin there is an 8 centimetre layer of blubber. All these features prevent heat from escaping from the bear's body. The polar bear has another way of keeping its body temperature steady. It curls up, when it is cold, making its surface smaller, or sprawls with its legs stretched out when it is too hot.

Polar bears live alone except for a few days when pairs come together to mate. During winter darkness they shelter, the females digging dens in the snow. Lying inactive in her den the female is protected (like a jerboa in the desert — p. 54) from the most severe winter weather. Here she has her young, usually twins, which are ready to leave by the time lighter days return.

SUMMARY

Some plants and animals live in large numbers in the harsh conditions. Food-webs link the plants and animals of one area together. Anything which alters the numbers of one species will affect some of the others. Each species has a pattern of behaviour which allows it to survive, find food and reproduce sucessfully if not disturbed and if conditions remain unchanged. Within any group of animals the larger species are usually found nearer the poles. They have less surface (through which heat can be lost) compared with their volume.

Arctic fox

Huddling of emperor penguins

Male emperor
penguins huddle
together while they
incubate their eggs
in the Antarctic
winter

Emperor penguins breed in very cold, windy places. The males huddle together (see photo) as they incubate their eggs. Huddling is supposed to keep the penguins from losing too much heat. Is this true?

Plan an experiment to test the suggestion. Use any of the apparatus shown in the drawing. You may not wish to use everything shown. You can use a tin wrapped in cloth and filled with warm water instead of each penguin. The water will be the warm inside of the penguin's body. The cloth will be the feathers. You can use the hairdryer to make the wind. Write down everything you would do and what results you would record. If possible, try your experiment and record your results so that you can test whether it does help the penguins to huddle together.

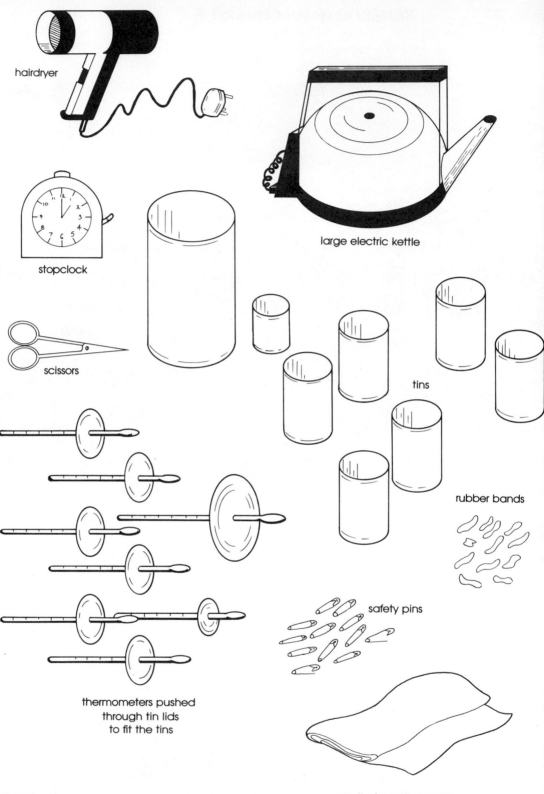

hairdryer

large electric kettle

stopclock

scissors

tins

rubber bands

thermometers pushed
through tin lids
to fit the tins

safety pins

a roll of woollen cloth

**Apparatus for the
experiment**

QUESTIONS ON CHAPTER 6

1 Write down each sentence, rearranging the mixed-up word to fill the gap.

(a) **HOPPYKLANNOTT** is the plant food on which all Antarctic animals depend.

(b) **LIRKL** is a small animal like a shrimp which is food for other Antarctic animals.

(c) Blubber, fur and feathers provide **TAILSUNION**, keeping birds and mammals warm in polar regions.

(d) The **LABSTORAS** is a carnivorous bird found in the Antarctic.

(e) The **DOLAPER** seal feeds on other Antarctic seals and penguins.

(f) The different kinds of Antarctic seals survive in different ways; they occupy different **CHEINS**.

2 Make a drawing of a food-web to show the information below. Give your drawing a title.
Information:

(a) Phytoplankton is the main plant food in the Arctic.

(b) Fish feed on phytoplankton.

(c) Seals eat fish.

(d) Polar bears feed on seals and fish.

(e) Humans feed on fish, polar bears and seals.

3 Make a list of animals which feed on krill. Write down what you think might happen to all these animals if too many krill were collected for human food. What other effects might there be on the food-chain?

4 (a) Explain what conservation of plants and animals means.

(b) Why is conservation important?

(c) Describe three examples of conservation.

(d) Make a poster or display about wildlife conservation.

CROSSWORD SOLUTION

Across

1 Predators
4 Men
6 Vulture
9 Gnu
10 Tree
11 Lioness
12 Cheetah
15 Eland
16 Rhinos
18 Cellulose

Down

2 Eat
3 Tsetse
5 Eggs
6 Vulnerable
7 Leopard
8 Seeds
13 Canine
14 Grass
15 Eyes
17 Sun

ANSWERS TO WORDFINDER:

1 Dormant
2 Camel
3 Eat
4 Arid
5 Oasis
6 Jerboa
7 Hump
8 Brine
9 Gregarious
10 Nomad
11 Camouflage
12 Cactus
13 Ear

14 Evader
15 Acacia
16 Dry
17 Eyes
18 Dromedary
19 Stoma
20 Desert
21 Famine
22 Corpse
23 Sand
24 Xerophyte
25 Fennec
26 Seed

INDEX